FIGHT GLOBAL WARMING NOW

FIGHT GLOBAL WARMING → NOW

THE HANDBOOK FOR TAKING ACTION IN YOUR COMMUNITY

BILL McKIBBEN

AND THE STEP IT UP TEAM—
PHIL ARONEANU, WILL BATES, MAY BOEVE,
JAMIE HENN, JEREMY OSBORN, AND JON WARNOW

A HOLT PAPERBACK   HENRY HOLT AND COMPANY   NEW YORK

Holt Paperbacks
Henry Holt and Company, LLC
Publishers since 1866
175 Fifth Avenue
New York, New York 10010
www.henryholt.com

A Holt Paperback® and ® are registered trademarks of
Henry Holt and Company, LLC.

Library of Congress Cataloging-in-Publication Data
McKibben, Bill.
    Fight global warming now : the handbook for taking action in
your community / Bill McKibben and the Step It Up Team,
Phil Aroneanu . . . [et al.]. — 1st Holt paperbacks ed.
        p.   cm.—(A Holt paperback)
    Includes index.
    ISBN-13: 978-0-8050-8704-8
    ISBN-10: 0-8050-8704-4
    1. National Day of Climate Action, 2007.   2. Green movement—United
States.   3. Global warming.   4. Carbon dioxide—Environmental
aspects.   I. Step It Up Team.   II. Title.
    QC981.8.G56M35 2007
    363.738'74—dc22   2007025492

Henry Holt books are available for special promotions and
premiums. For details contact: Director, Special Markets.

First Edition 2007

Designed by Kelly S. Too

Printed in the United States of America
♻ Printed on recycled paper
1   3   5   7   9   10   8   6   4   2

For the 1,400 organizing teams across America
that made April 14, 2007, such an exciting day

CONTENTS

STEP IT UP

It's easy to join the global warming movement. We know it's easy because we all just joined ourselves. None of us has spent long years as organizers. One of us has spent long years mostly as a writer with a little activism on the side; the rest of us haven't spent long years doing anything except school, because we just got out of college.

But in 2007 we came together to see if we could kick up a fuss about climate change. That January 10, we launched a Web site, StepItUp2007.org. We asked people across the country to start organizing rallies for April 14, to demand that Congress cut carbon emissions 80 percent by 2050. We had no money, and we had no organization, so we had no expectations. Our secret hope, which seemed a little grandiose, was that we might organize a hundred demonstrations for that Saturday, only three months away.

Instead, our idea took off. The e-mails we sent ended up spreading virally, in the way that certain ideas sometimes do on the Internet. People we'd never heard of started signing up on the Web site to host rallies in places we'd never heard of. The electronic pins stuck on our online map got thicker by the week—200, 500, 900. By the time the big day rolled around, there were 1,400 demonstrations in all fifty states, ranging from tiny to enormous. It was one of the biggest days of grassroots environmental protest since the first Earth Day in 1970, covered extensively in the national media and in thousands of local stories across the country.

Along the way we learned a few lessons and we want to share them in this book, which is designed to help you plan and carry out your own ongoing local rallies and campaigns, the way the thousands of organizers we worked with did on April 14, 2007. We agreed to write it because, one, we haven't quite managed to solve global warming yet and, two, we gained a few hard-earned ideas for how to make the most of two things: local communities and the Internet.

There is no shortage of fine books on activism, from Saul Alinsky's classic *Rules for Radicals* through much more recent accounts. Many of them have centered on the very difficult, long-term, and noble task of community organizing—convincing people with too little power to stand up for their rights. We're mostly talking about something a little simpler here: getting Americans who already care about an issue such as global warming to actually take effective political action. And we think certain things about contemporary America offer both opportunities and pitfalls for organizers. This isn't the 1960s anymore; an awful lot has changed, even in the last few years.

We had an excellent database to draw on: all the people

who organized events for Step It Up and then sent us pictures and reports. We interviewed and surveyed a great many of these organizers to learn what worked and what didn't, and this book is as much their work as ours. (That's one reason why any proceeds we receive from sales of the book will go back into the climate change movement.) We also drew on our work organizing the biggest climate change protest march to date, in the summer of 2006, as well as various campus campaigns we've been involved in. We wrote this book as if you were getting ready to organize a rally in your community and want practical help thinking it through and pulling it off. For "rally," you can substitute a lot of other ideas—teach-ins, petition drives, phone-banking, voter-registration drives. Any kind of action or campaign, we think, can benefit from at least some of these tips.

We also think that our experiences offer insights for those working in other social change movements beyond the environment. We'll be illustrating our points with examples from our experience, but you'll be able to see pretty easily how they might fit other causes. We'll provide detailed, nuts-and-bolts advice, but we're grouping most of our thoughts more thematically, because we've found these organizing principles to be powerful.

- **Make it credible.** You need to know enough about your subject to argue convincingly, but you certainly don't need to know everything, and you shouldn't be intimidated by the fact that you're not an expert.

- **Make it snappy.** Today, it's easier to organize ad hoc actions on short notice (thanks mostly to the Internet) and harder to get people to join organizations, come to endless meetings, and so forth (thanks to longer work days, commutes, and the like). So we describe the benefits of short-term campaigning for making your point.

- **Make it collaborative.** In an age when our leaders are often hopelessly split along partisan lines, it's actually quite possible, and quite necessary, to reach out to diverse kinds of people to make a stand against something as all-encompassing as global warming. We think *sharing* an action instead of *owning* it is key.

- **Make it meaningful.** People are eager for the chance to do something that shows their real commitment—say, walk for a day (or even a week). In a religious nation, many are eager for the chance to put faith to public use and to take a stand in the places that matter most to them. Moral seriousness makes an important impression.

- **Make it creative (and fun!).** Along with earnestness, we have found that the best actions are fun to do and fun for others to consider. The environmental movement hasn't been much of a singing movement for years; art has played too small a role. We describe some ways to involve everyone, from actors to athletes.

- **Make it wired.** Activism can't live solely on the Web—virtual petitions and the like aren't that powerful. We do think, however, that the Internet has become the crucial tool for building momentum behind the kind of actions that can fight global warming, and we think there are some things to understand as you put it to use.

- **Make it seductive (to the media).** A successful action doesn't require any coverage at all. But it's easy to amplify the effect of your hard work if you can get reporters and editors to pay attention. And that's relatively easy to do if you understand how they think.

- **Make it last.** Ad hoc organizing can lead to future actions, and the nature of working together in the short term often builds long-term bonds.

We draw largely on our experience with Step It Up 2007 in this book, but we've worked on other actions, some of them successful and some not, that have taught us lessons, too. Organizing is extremely interesting work. (Well, most of the time. Sometimes it's just filing for permits and waiting in line at Kinko's.) It's as much about human nature as it is about political strategy, as much about the small issues of how we relate to one another as it is about the big issues of the day.

In his book *Blessed Unrest,* our friend Paul Hawken said that the movement that is rising to stop global warming and many other planetary inequities will be the largest our planet has ever seen. We want to give you the tools to ensure he's right.

Only three years ago, global warming was off the radar screen for many Americans. Today, it is in the national spotlight and a diverse network of groups is rising to the challenge of stopping it. Hundreds of colleges and universities are working to become carbon neutral, reducing emissions from campuses to zero. Community organizers in Oakland, New Orleans, Detroit, and elsewhere are taking on polluters and fighting for environmental justice. In Appalachia, rural communities are banding together to fight mountaintop removal, a heartbreaking new method for mining coal from that region. People of faith are organizing their churches, synagogues, and mosques, declaring global warming as the moral crisis of our time. Traditional businesses are greening up, while entrepreneurs are building a clean-energy alternative economy that has the potential to create thousands of new jobs. And this is just the beginning.

Despite the array of groups and organizations working on global warming, we are still missing a key element: the *movement*. Along with the hard work of not-for-profit lobbyists, environmental lawyers, green economists, sustainability-minded engineers, and forward-thinking entrepreneurs, it's going to take the inspired political involvement of millions of Americans to get our country on track to solving this problem. Linked up by the Internet and a common vision, we can start to make change from the local level to the national and global. We hope this book will give you the skills and inspiration you need to jump into this growing movement. It's hard work, but—take it from us—it can be a lot of fun, too.

In 1968, observing the state of civil rights in America, Rev. Martin Luther King Jr. said, "We are now faced with the fact, my friends, that tomorrow is today. We are confronted with the fierce urgency of now." Today, we are feeling that fierce urgency again for two reasons. The first is that scientists are telling us that we are running out of time even faster than we thought. If we don't act within the next few years, we won't be able to avoid the worst effects of climate change. The second reason is a more hopeful one. Recent political changes in Washington DC and around the country have finally created an opportunity for genuine political action on global warming. There is no guarantee that this situation will last. If you've been a little paralyzed by the sheer size and horror of global warming, now is the time to start moving forward, *fast*.

FIGHT GLOBAL WARMING NOW

‣ 1 ‣

MAKE IT CREDIBLE

If people know that something is wrong and dangerous, why don't they organize to do something about it?

One of the most common reasons is the sense that they don't know enough—they have strong points of view, but since they're not experts they fear they won't be able to talk about it clearly or answer every objection. When something seems "scientific" and "complicated" like global warming, those misgivings can be especially strong.

This is understandable. It's also wrong. If you want to do original scientific research about climate change, then you better head to graduate school in chemistry or physics or biology. But if you want to talk about it—to your neighbors, to the media, to your political leaders—then there's no need to know absolutely everything. You just need to be able to speak strongly about the essentials of the issue.

This is true as well about every other issue you can name. You don't have to be an economist to argue for a living wage or a general to address issues of war and peace. In fact, most people wouldn't be frightened away from those topics since they believe in their expertise as workers and as citizens. Unfortunately, when it comes to the science of global warming, it has been easy to confuse the public by throwing around "competing" statistics. So though the most important things you bring to the movement are your passion, your wit, and your energy, you do have to know a little, if only to make the case for organizing against climate change with confidence.

How much information do you need? We have pulled together this primer as a series of answers to the kinds of questions most often thrown at climate change activists.

WHAT EXACTLY IS GLOBAL WARMING?

When we burn fossil fuels, we emit carbon dioxide into the atmosphere. A *lot* of carbon dioxide—a gallon of gas weighs about eight pounds, and when it's burned in any kind of engine, about five and a half pounds of carbon dioxide (CO_2) goes out the exhaust pipe. And there's no filter you can put on your exhaust pipe, or most other sources of carbon dioxide, to reduce that flow. Since carbon dioxide is an inevitable by-product of fossil fuel combustion, the only way to reduce it is to burn less coal and gas and oil.

It's important to reduce carbon dioxide because its molecular structure traps heat that would otherwise radiate back out to space. It's like an invisible blanket in the atmosphere, or the panes of a greenhouse. There's always been some carbon in the atmosphere, which is a good thing—without it, the world would get very cold. But ever since the start of the

Industrial Revolution, when we began to burn fossil fuels in large quantities, the amount has been increasing. There's more of it in the atmosphere now than there has been for millions of years.

Scientists tell us we have so far raised the average temperature of the planet about one degree Fahrenheit, from roughly fifty-nine degrees to sixty degrees. The strong scientific consensus is that unless we act very quickly and powerfully to reduce the amount of fossil fuel we're burning, we will raise the temperature another four to eight degrees in the course of this century.

YEAH, BUT SO WHAT? ONE DEGREE DOESN'T SOUND LIKE MUCH. AND I LIKE IT WHEN IT GETS WARMER.

One degree *doesn't* sound like much. But Earth turns out to be a finely balanced place, and already we can see the effects of even that small amount of warming. Everything frozen on Earth is melting—new data show that Mount Kilimanjaro may lose its snowcap inside of a decade. The seasons are changing fast. Scientists say we have both more drought and more flooding. (Drought because warm air holds more water vapor—there's increased evaporation in warmer areas—and flooding because once that water is up in the clouds it's eventually going to come down.) Hurricanes are growing stronger and lasting longer.

And that's with one degree. Five degrees more would make Earth warmer than it's been since long before humans arrived on the scene. In the words of the eminent NASA scientist James Hansen, it would create a "totally different planet." We don't know exactly what this planet would be like, but all the world's leading climatologists and earth scientists have joined together to give us a pretty good idea. They formed a body called the Intergovernmental Panel on Climate Change (IPCC), which issues a new assessment every five years of everything we know about global warming from the peer-reviewed scientific literature. Their latest set of reports was issued in the winter of 2007.

Those reports, though written in the dry language of science, tell us many things about that future. Global warming will cause an increase in human deaths, as mosquito-borne diseases like malaria spread more widely. There will be huge increases in flooding and severe droughts across wide areas of Earth. A large percentage of the world's plants and animals—as many as 40 percent—will move closer to extinction, or over the edge altogether. The changes are so large it's almost hard to imagine them: an Arctic without any summer ice, for instance. And yet the data tell us that ice may disappear as early as 2020.

DO WE KNOW EXACTLY HOW MANY PEOPLE WILL DIE AS A RESULT, OR HOW MANY SPECIES WILL BE WIPED OUT?

No, of course not. One of the worst things about the changes we're making to Earth is that we don't know exactly where they'll lead. It's a huge experiment, the largest thing humans have ever done to the planet.

On April 14, 2007, the Step It Up national day of action on climate change, thousands gathered outside of the Capitol Building in Washington, DC, to create a "human postcard" asking Congress to cut carbon emissions "80% by 2050." (PHOTOGRAPH BY JOHN QUIGLEY/ SPECTRAL Q)

That's one of the reasons that some of the first nonscientists to get really worried about climate change hailed from the insurance industry, the people charged with analyzing risk in the economy. A recent study sponsored by the world's largest insurance firm, Swiss Re, and Harvard Medical School predicted big increases in diseases such as malaria, higher risk of crop failures, and repeated devastation from floods and storms. It concluded that such changes will make economic growth in poor countries steadily more difficult, and that even in affluent countries like the United States many regions could "experience developing nation conditions for prolonged periods as a result of natural catastrophes." And one disaster piles on top of another, so pretty soon it's hard to recover. New Orleans post–Hurricane Katrina appears to be our first real taste of this phenomenon.

COULDN'T THIS JUST BE A NATURAL CYCLE? AFTER ALL, WEATHER DOES CHANGE ALL THE TIME.

Weather does change all the time, and so the first question that scientists tackled when they started worrying about global warming was whether the current heating fell outside the normal range. It took them a little while to analyze all the data, but by the mid-1990s the scientific consensus was that there was nothing natural about this new heat—the only possible cause for most of it was the carbon we were pouring into the atmosphere.

That confidence has grown stronger with each new year of scientific research. In May 2007, IPCC's new five-year report stated a scientific consensus, based on tens of thousands of studies: there is virtually no possibility that the heating we're seeing can be explained by anything other than human causes.

WHERE WILL THESE CHANGES BE FELT FIRST?

The impact of global warming is already being felt around the world. Arctic cultures, from the world of the Inuit to the world of the polar bear, are being turned upside down— villages have been evacuated and old hunting traditions are disappearing. For a number of technical reasons, the planet's temperature rises most quickly at the poles.

Over time, there will be even larger effects in the most populated parts of the world—the coastal plains of the tropics. Billions of people live on and depend on food from land near the sea. Many will simply have to leave—the United Nations predicts hundreds of millions of refugees by midcentury. But of course if a person is forced to leave Bangladesh, it's not at all clear where he or she will move to—there's not a lot of empty land in the neighborhood. Already some small island nations have begun relocating residents to higher land in nearby countries, including New Zealand.

For people concerned with social justice, this is especially sad news. People in these tropical countries are among the poorest in the world, and they have done nothing to create this problem—the 140 million people in Bangladesh hardly produce enough carbon dioxide to measure.

Even close to home, poor people and minority communities in coastal areas are bearing the brunt of the warming planet (think again of Hurricane Katrina). This is one of the reasons that many religious and human rights groups have become involved in efforts to fight global warming. Environmental justice advocates are working side by side with traditional environmental, religious, and youth groups to ensure that the coming green tide is, as our friend Van Jones puts it, "strong enough to lift all boats."

WILL ANYTHING HAPPEN WHERE I LIVE?

Yes. In recent years, scientists have developed models that can better predict the localized effects of climate change. Most parts of the United States have been studied, and the results are disturbing. In New England, for instance, the models show that over the course of the century, winter will become shorter and shorter and eventually disappear altogether. The forests of birch, beech, and maple that turn such glorious colors in the fall will not be able to reproduce in the warmer temperatures. We won't even have maple syrup season in the spring.

In the Pacific Northwest and California, winter snow-packs are expected to become much smaller—which means much less water being stored behind dams for use in the dry summers. Water will also be a problem in the Southwest, where warmer temperatures will mean less flow in rivers like the Colorado. Heat waves across the Great Plains may cause big trouble for agriculture, and storms are on the rise. In the Southeast, exposure to fiercer and more frequent hurricanes will exact a toll. Disease vectors like ticks and mosquitoes will spread steadily farther north. And if you live near the coast, water will rise ever higher. Everywhere extreme heat will become more likely and, with it, the danger of wildfires.

THE LOCAL EFFECT

Bring the climate change issue home for your community.

- Find a local climate change expert who can discuss the predicted effects of global warming on your region or community. Your local college or university is a good place to start.
- What specific changes will most affect the daily life of your neighbors? If it's flooding, ask how high the coastal waters will likely rise or how many more hurricanes are likely to hit. If it's hot-weather droughts, ask how much less water will be in the region's aquifer. The local expert won't answer with specific numbers but may be able to give you a range.
- Always ask if you may call on the expert as a speaker, guest writer, or media source in the future.

WILL COASTAL CITIES REALLY BE SUBMERGED?

The short answer is, we don't know. Sea levels rise for a couple of reasons. One is because warmer water takes up more space than cold. This effect alone should cause sea levels to rise at least a foot this century.

But sea levels will also rise if the great ice sheets above Greenland and of the West Antarctic begin to melt. Scientists used to assume this would take a very long time. Now many are starting to say it could happen much more quickly, because melting ice is falling to the bottom of these ice sheets and greasing the skids for their slide into the ocean. If that happens, it's bad news; sea levels could easily rise twenty feet by the end of the twenty-first century.

Janisse Ray, author of *Ecology of a Cracker Childhood*; her husband, Raven Burchard; and their friends "water-tubed" Vermont's West River in January—a month when it would normally be covered with two feet of ice—to give an early publicity boost to Step It Up 2007. (PHOTOGRAPH BY SAM KILMURRAY)

This is the most controversial part of climate forecasting. The rest—the five-degree rise, the flooding and drought, the spread of mosquitoes, the death of trees, the end of winter—are widely accepted as *likely* to happen. Let's hope the scientists discover that these ice sheets won't melt, because it's one more horror we don't need, on top of everything else.

HOW CAN WE STOP ALL THIS?

Unfortunately, we can't stop all of the effects of global warming. The one degree the world has already warmed is here to stay, and scientists think we have already released enough carbon into the atmosphere to add another degree and a half to global temperature. But that's still a lot less than the four-to-eight-degree increase we will be facing if we don't act dramatically—and the difference between a difficult future and a catastrophic one.

The IPCC—the international group of climatologists—said in May 2007 that if we act decisively now we should be able to hold temperature increases below the level at which the ice sheets are guaranteed to melt. If we act quickly, we may be able to shave two or three degrees off Earth's eventual temperature and slow down the heating of the planet.

Half measures won't do the trick. Occasionally, opponents of action on global warming say that even if the United States had joined the Kyoto treaty, global warming would have been delayed only a little bit. This is true: designed in the 1990s, the Kyoto treaty was an unambitious first step. Since then, we have learned a great deal more about how to slow global warming and developed new strategies and technologies. The scientists tell us that in America we need a plan to cut carbon emissions at least 80 percent by 2050—that's why we made it our message for Step It Up.

BUT AMERICA CAN'T DO THIS BY ITSELF,
CAN IT? WHAT ABOUT CHINA AND
INDIA AND THE REST OF THE WORLD?

Indeed, we can't solve global warming without the participation of the fast-growing developing world. But if we're going to get that cooperation we need to realize the basic facts.

The United States has been and still is the biggest source of carbon dioxide. The 4 percent of humans who live in America produce nearly a quarter of the world's CO_2. Perhaps as early as this year, China will pass the United States as the largest annual emitter of carbon dioxide, but when it does China's per capita emissions will be just a quarter as large as America's. (Per capita emissions are the only way to think about solving this problem—otherwise, a country like China could "reduce" its emissions just by dividing itself into four separate nations.) Not only that, but carbon dioxide stays in the atmosphere a long time, which means that Americans' many decades of CO_2 production will cause most of the world's problem for many years to come; it may be forty years before China is responsible for as much of the global warming problem as the United States, and even then each Chinese resident will be only one-quarter as much to blame as each American.

So if U.S. leaders are going to persuade the Chinese, the Indians, and others to help solve the climate problem, they are going to have to approach them a little humbly and with the offer to transfer the technology necessary to develop without adding to global warming. There are signs that these countries might be willing to strike a deal—China, for instance, has stiffer mileage standards for its cars than the United States does—but we have to get back to the negotiating table. Think about how Americans appear to the rest of the world—producing more carbon dioxide than other people and then letting our leaders ignore the problem. Until we come up with a credible plan for taking care of our own emissions, no other country is going to give us the time of day. That's one of the biggest reasons it's so important for Washington to act now.

WILL IT COST TOO MUCH
TO DEAL WITH THIS PROBLEM?

The price tag is one reason some American leaders have refused to take serious action on global warming. The George W. Bush administration has maintained that it would seriously harm the U.S. economy if the government enforced reduced carbon emissions. Opponents claim it could cost hundreds of thousands of jobs or wipe out whole industries.

If we take cutting carbon emissions seriously, it *will* damage certain industries—coal mining, for one, since coal is the fuel that produces the most carbon when it's burned. But the number of coal miners is relatively small, and they can (and should) be retrained for other work—including repairing the horrible damage that mountaintop-removal mining has caused in their own southern Appalachians. As some economic sectors are hurt, however, others will flourish: many Americans will find jobs building or installing solar panels, manufacturing new energy-efficient technologies, or establishing new, more local methods of distribution for commodities such as food, which often travels thousands of miles.

Experts can estimate the total effect on the economy using a variety of models. The biggest one, using the most recent data, was published in May 2007 by the IPCC. The report—the work of 168 authors, 84 coauthors, and 485 peer reviewers from a huge variety of fields—found that making the cuts necessary to hold temperature rise short of the catastrophic range would cost about one-tenth of 1 percent of the world's gross national product each year. That would mean reducing the size of the world economy just 3 percent by 2030. We would have to wait until Thanksgiving 2030 to be as wealthy as we would otherwise be on New Year's Day 2030, but in return we would have transformed the energy system.

THEN WHY IS BUSINESS SO OPPOSED TO ACTION?

Business isn't opposed to action. One company after another has shown that it can cut carbon emissions and save money in the process. Even some very conservative companies, such as Rupert Murdoch's News Corporation, owner of the Fox News Channel, have announced that they are going carbon neutral.

Predictions of misery and woe generally come from those industries with something to lose. ExxonMobil, for example, has spent huge sums of money to convince Americans they don't need to change, and that it would be ruinous to do so. But such vested interests always predict that change will be too expensive—the auto companies made that claim about seat belts and then about air bags. Experience proves that once we get started, our ingenuity will let us make changes more quickly and cheaply than we imagined.

And if we don't get started, others will. Over the past decade, Japan and Germany have taken solar power much more seriously than the United States, providing government subsidies to get the industry rolling. They have since been able to lower and end those initial subsidies—and, unsurprisingly, those countries now host most of the world's solar panel factories. If we keep propping up yesterday's industries, we'll end up with yesterday's economy.

This helps explain why so many local political leaders are intrigued by the prospects of a low-carbon economy. As California's Republican governor Arnold Schwarzenegger explained in the spring of 2007, new global warming standards set by the state have spurred business, not harmed it: "We have seen so many companies that have been created that work just on things that have to do with clean environment. There are a lot of industries that are exploding right now because of setting these new standards."

THE BUSINESS ANGLE

If you aren't a business leader yourself, you'll want to take advantage of the companies that are going carbon neutral.

PRACTICAL TIP

- Research the businesses in your town or state that provide clean energy options or are making noise about their efforts to be ahead of the curve on addressing global warming.

- Involve local low-carbon business leaders. They may be able to provide "in-kind" donations (recycled paper, photocopier time) or loan equipment (camcorders, sound systems); provide a meeting or event space; and pull people in through local business associations. More important, they could provide credible spokespeople to show it's not "just environmentalists" who care.
- Ask business leaders to work with politicians and their lobbying organizations in Washington.

WILL OUR LIFESTYLES HAVE TO CHANGE ENORMOUSLY?

Probably not. Most of the first cheap and easy steps won't involve enormous change. We'd need to raise fuel efficiency quickly and dramatically, which would mean vehicles designed for forest rangers would become much less common in suburban parking lots. But new technologies like hybrid engines can provide mobility with half or a third as much gas (and half or a third as much pain at the pump). New compact fluorescent lightbulbs provide as much light as the incandescent ones they replace—and last much longer and cost much less to operate.

Over time, plenty of things will change—and many of them for the better. Instead of flying or trucking the average bite of food fifteen hundred miles, for instance, more Americans will find themselves eating locally. In fact, this is already starting to happen: farmers' markets are the fastest growing part of the U.S. food economy, and, according to surveys, customers at green markets are far more satisfied than super-market shoppers. Instead of sprawling ever farther out into the countryside, we will need to start filling in some of the vacant spaces in our cities and suburbs, creating the kind of clustered and close communities that many Americans say they prefer over our current subdivisions. Instead of relying on energy from far away—energy we must try to safeguard with American lives, at enormous expense—we will need to start producing more of our own. Already lots of people have put solar panels on their roofs, and when the sun is out they feed electrons into the grid, functioning as small utilities (while their electric meters spin the other way). That kind of energy system is a lot like the Internet—people put in, people take out. And it offers the promise of a more durable as well as a more ecological energy future.

People who have traveled to western Europe have already seen a different energy future. Germans or Italians or Britons lead perfectly decent lives (in fact, surveys show they're more satisfied with their lives than Americans are). But because they have slightly different habits, like the willingness to take public transit when it's available, they use *half* as much energy per person as Americans. Half is a lot.

WON'T WE NEED TO BUILD LOTS AND LOTS OF NUCLEAR PLANTS?

Probably not. Although nuclear plants produce relatively low-carbon electricity, they burn something else: money. Given the $3 billion or so it takes to build a new reactor, it's easy to find other ways to spend the money that will reduce carbon emissions much faster. (Conservation, for instance, is much cheaper, and since we use so much energy so wastefully, there will be lots of energy to save for years to come.)

The IPCC estimates that worldwide we would need only a 2 percent increase in the share of our power generated by nuclear plants to reach low-carbon targets, and other experts say even that may not be necessary. The American Solar Energy Society, for instance, has published a plan for generating half our power by 2030 from renewable energy alone.

But the possibility of nuclear power does allow us to understand something else: the true danger of new coal-fired power plants. An atomic reactor, as we all know, comes with

More than three hundred residents from Ames, Iowa, marched and rallied in front of the city's coal plant in a call to reduce carbon emissions. (PHOTOGRAPH BY GREG BAL)

risks. But a new coal-fired power plant comes with a *guarantee* of destruction. If any significant portion of the coal plants now on the books in this country are built, we will miss any chance for dealing with global warming. The answer is not to trade them in for costly new nuclear reactors; the answer is to get to work on conservation and on renewable power.

WHAT ARE THE FIRST THINGS WE NEED TO DO?

As of right now—2007—the list of priorities is clear. Americans need to get Congress and the president to pledge to cut carbon emissions at least 80 percent by 2050. This long-term goal will signal to investors and planners that they need to put their money and creativity into technologies and buildings that make sense in a new energy future.

We need interim steps too, to make sure that politicians don't keep pawning off real change on their successors. A modest proposal would look like this:

- An immediate moratorium on new coal-fired power plants.
- A freeze on carbon emissions by 2010. Before we can start getting emissions to go down, we've got to stop them from going up.
- A commitment to 30 percent carbon emission reductions by 2020, most of which will come through conservation, increased gas mileage, and the like.
- A commitment to having the country produce 25 percent of its power from renewable sources such as the sun and the wind by 2025.

- A government Clean Energy Jobs Corps program that provides access to jobs in order to jump-start the retooling of our country's energy system.
- A real commitment to international efforts to invest in a rapid and fair transition to a clean energy future.

WOULDN'T IT BE EASIER TO GET THESE GOALS APPROVED AT THE STATE AND LOCAL LEVELS?

Because Washington has refused to act for so long, people have applied useful pressure to city halls and state legislatures—and gotten impressive results. Hundreds of cities have joined climate initiatives; California has pioneered impressive legislation. This work has produced much knowledge and many great examples to build on, and it must continue—ultimately, the real change will have to happen in all the places we actually live and work.

But only Washington can change the price of energy and send a signal throughout the economy. And only Washington can credibly negotiate with the rest of the world to reach an international agreement. That's why climate activists are aiming a lot of their work at the federal government right now—and there seems to be a good chance of success, since more and more politicians are waking up to the problem.

There's plenty of good organizing to be done closer to home, too—and not just directed at the government. People are persuading their places of worship, their colleges, and their retirement communities to go carbon neutral. Pay attention—

some months openings for action are closer to home; at other times, especially as elections approach, it may be easier to influence Washington. Try it all!

WOULDN'T THE BEST PLACE TO START BE AT HOME, WITH INDIVIDUAL ACTIONS?

Indeed it would. There are many changes we can make easily at home, like that new lightbulb over the kitchen table. But organizer after organizer has told us that they became active because, even as they were screwing in the new lightbulb, they realized such action by itself was unlikely to address the emergency we're in. They did the math—the change we need is so sweeping and so rapid that only by mobilizing ourselves through our government will we be able to make enough progress in the time we have left. Another unofficial motto at Step It Up: "Screw in a lightbulb, and then screw in a new federal energy law."

STARTING AT HOME

The home front alone won't possibly win the fight against global warming, but if someone asks what they *can* do, you can offer seven simple actions.

- Replace common incandescent lightbulbs with low-energy compact fluorescent bulbs (CFLs). (By the way, of the ideas on this list, this one will probably have the least impact on your personal carbon budget, saving about a hundred pounds of carbon per lightbulb per year.)

PRACTICAL TIP

- Get a programmable thermostat and adjust it two degrees cooler in the winter and two degrees warmer in the summer. Even better—skip the air-conditioning when it's hot and wear an extra sweater when it's cold.

- Keep your water heater set no higher than 120 degrees.

- Purchase an energy-saving (Energy Star) washing machine or refrigerator or other appliance when you replace your old one.

- Switch to green power by purchasing renewable energy from your local utility via the Climate Counts Web site (www.climatecounts.org/get_clean.html).

- Buy products such as groceries from local sources, and walk or use mass transit to get to the market.

- If you have to drive a car, keep your tires properly inflated. Combine trips that require an automobile, and take the vehicle that gets the best gas mileage. And make it a hybrid and carpool!

There are lots of similar action steps listed on our friend Laurie David's Web site (www.StopGlobalWarming.com). And though small steps won't be enough, an organized lightbulb exchange or other efforts can be an appealing way to pull first-time activists into joining your planning team and the larger global warming movement.

WILL THESE ACTIONS BE ENOUGH?

At the moment, the best science tells us that these steps, if enacted quickly, would allow us to slip through a very narrow window and avoid complete catastrophe. But that window is closing—we have to keep monitoring new research, because year after year it shows that the planet is reacting more and more quickly to the warming we have touched off.

No one can guarantee that these steps will be enough. But we can definitely guarantee that not taking them will cause grievous damage. And following this path will not just help ward off climate change—it will also help us build a more resilient and just society, less dependent on faraway sources of energy.

These answers will deal with most of the questions that curious, normal people will have for you. But you may encounter some of the dwindling band of climate skeptics, people who have convinced themselves that it's getting warmer because of something beyond our control (sunspots are a common candidate) and that it will soon begin to cool again naturally. It's unlikely you'll actually convince such people, who often have some kind of deep-seated political agenda or ideological tilt. And you don't need to convince them. Political action doesn't demand winning over everyone—just enough of them to force change.

Two organizers in Davis, California—fourteen-year-old Shealyn Wallace and sixteen-year-old Alex Wallace—have a pretty good outlook on the whole thing: "Some drunk college guys [at our event] tried to yell that global warming is a myth and we are brainwashed, but we talked to them and asked them to be respectful. They went away with their embarrassed girlfriends; they will be in trouble later with them so we thought it was fair punishment for being rude. Every-one else came and listened and cheered."

Our way of dealing with naysayers is first to thank them for their interest in the problem. Then we remind them that the scientific consensus on this issue is extremely strong, and has grown steadily stronger—there is essentially no peer-reviewed research in recent years challenging the basic science of global warming, only a small band of skeptics, many of whom have been funded by the fossil fuel industry. Then we say that we sure hope they turn out to be right be-cause there's nothing we'd rather see than an end to global warming. Yet because the science is so clear, we think it makes more sense to pay attention to it and take the feasi-ble, economically sensible steps the scientists tell us we need to take. After that, it's time to get back to the business of organizing.

2

MAKE IT SNAPPY

Just the word—*organizing*—makes it sound like something that takes a long time and involves a high degree of sophistication. And that can be the case. It can take years and years to turn a group of isolated people into a unified force, like a union.

But we're talking about something a little different here—convincing other people who already know that something is wrong to do something about it. And we think this work needn't take forever—indeed, that it's often improved by speed and urgency.

Here's the backstory to Step It Up. A group of Middlebury College students taking a class on climate change started meeting outside the classroom to talk about how they could take action on the issue. We met in dining halls, in dorm rooms, over breakfast in the morning, and over beer at night. It got to the point that many friends began to wonder what the hell we were doing all the time, and wanted to voice their own thoughts and ideas in the discussion. The group became larger and, as class schedules got in the way, we began to meet every Sunday night, calling our ad hoc action forum the Sunday Night Group. That was in February 2005, and many of us writing this book have been hanging out in those meetings ever since. We got to know one another there, sharing passion, energy, organizing tips, and the ups and downs of campaigns, food, song, and friendship.

Then one day in early August 2006, Bill decided that Something Had to Be Done to spur the larger activist movement on global warming in the United States. Though he'd written the first mainstream book on global warming and spent many years speaking about the issue, he was somewhat clueless about spurring activist movements, so his grand plan involved gathering a few friends, walking across his home state of Vermont, and staging a sit-in on the steps of the Burlington Federal Building. The goal would be to get arrested, make some headlines, and, we hoped, get people riled up enough to get out and do some more. It was a fairly dubious plan, despite the passion and enthusiasm behind it. At some point, someone had the good sense to check in with the Burlington Police Department, and it turned out that they were quite happy to let us sit on those steps as long as we liked. No arrests, no drama—no movement.

So Bill and the Sunday Night Group retooled. Over three weeks, we laid the groundwork for a mass march up the Champlain Valley of Vermont—a five-day trek that eventually drew a thousand people (which in Vermont is a lot). It took enormous work, with permits, campsites, food, loudspeakers, and musicians, but everything was in place in just *three weeks*. It was one of the biggest actions on climate change ever held. The most important lesson we learned: *You can do things more quickly than you might have imagined.*

The speed and success of the Vermont walk spurred us on, and within a few months we were talking about something bigger and better: Step It Up. A hallmark of the Step It Up day of action was that it too happened fast. We started talking about the idea around Thanksgiving and launched our Web site on January 10, so the most that any of the 1,400 organizers spent on the project was three months. Many spent much less time—some didn't even find out until a week or ten days before April 14 and they still managed to pull off credible events. We aren't *exactly* recommending doing things at the last minute. Disorganization is never an asset—but timeliness, liveliness, and flexibility are. Momentum depends on a sense of movement and urgency, and one of the best ways to create that sense of things is to make it snappy.

The prize for snappiest single Step It Up organizing drive may go to Doreen Simmonds. Doreen lives in the northernmost city in the entire United States: Barrow, Alaska. She found out about Step It Up on April 11, three days before it was set to happen. She writes, "I wondered if I could organize something in such a short time. To the universe, I said a quick 'Yes!' and proceeded to send out e-mails and make some signs." In just three days, she targeted Barrow's annual Spring Festival Parade, cleared a Step It Up marching delegation with city officials, and ended up at the front of the pack, declaring with her signage, "Congress: Give the Polar Bears a Break!" She even passed off another sign to the area's superior court judge, who was grateful that someone from Barrow had taken the leadership to call on Congress regarding global warming.

We also heard from people like Alice Bullock in Winston Salem, North Carolina. With two weeks until the big day, Alice learned that nothing was planned for Step It Up in her medium-sized city. There was little time to reflect or plan, and so she did what came naturally. She partnered with a local bike store that was hosting rides on April 14 and got the local Sierra Club office on board to make a banner and draft a press release. The day before, the public radio station interviewed Alice about the effort. During the event, Alice says she played the host, introducing people to one another and getting conversations going about global warming. People would sign her petition to Congress, and then stay to chat, watch the face painting, or listen to the impromptu jazz band of local high school students. The whole thing ended up on the front page of the local Sunday paper, a publication that had covered not a single story on global warming to date. Dozens of people organized events with two weeks or less, enough to embolden anyone to give activism a try.

Here's the take-home: we live in an age when the ad hoc and the impromptu are more possible than ever before (largely thanks to the Web). And speed confers all kinds of advantages, which we describe in this chapter. Most of all, having a deadline not too far in the distance adds a sense of urgency to everything you do. People spend less time infighting because they have less time to fight. And they pitch in because they have to—it's pretty clear that things need to get done.

Ecologically Conscientious Organization (ECO), a group of students from the University of Texas at Brownsville and Texas Southmost College, hosted speakers on various aspects of eco-activism. (PHOTO-GRAPH BY CARMEN GARCÍA)

SET THE TERMS OF DEBATE

It's amazing to watch the masters of the game play chess. Sometimes it's the very first move that determines the entire strategy, on both sides, for the rest of the match. Or consider pool—a player who makes a good break can sink two, three, four balls before an opponent even gets a shot off. It's no different in organizing. One of the best things about pushing the front line of an issue is that, well, you're the first one there, which means that you get the first crack at how everything is going to move forward.

If you define the issue wisely, it can give you a distinct advantage. With Step It Up, we chose to frame the debate around a slogan: "Step It Up, Congress: Cut Carbon 80% by 2050." That message worked to our advantage. We didn't have to spend much time explaining numerous policy goals; our one goal was right out in front. We also didn't have to deal a lot with the "climate change debate" of whether humans are, in fact, causing global warming. We wanted to move past that, so instead of focusing on the fact that climate change is real, we focused on the next step: what we need to do about it. And one result was that we got very few questions on the "truth" of climate change.

The other benefit of being out in front is that if you're fast enough, there's no time for adversaries to organize any substantial opposition. Step It Up was a ten-week campaign, a time line that made it difficult for a large special interest group to unroll a serious opposition. If you have ever worked in a large corporation, you probably know how long it can take for employees to schedule a meeting—weeks. And as a general rule, lying takes more time than telling the truth. Our slogan "80% by 2050" was also a hard message to reframe. If we had said, "Step It Up, Congress: Get Serious on Climate Change," it would have been easy for members of Congress to declare they were committed, say, to voluntary business carbon reductions as a serious step forward in reducing our greenhouse gases. A call for 80 percent reductions in carbon emissions by 2050 made it very hard for them to respond with anything less than a "yes" or a "no" supporting a bill with that benchmark.

When you begin to organize your action, see if there's a particular part of the movement in your area that needs pushing—then get it out front, and make sure your messaging sticks.

BE HIP

In his book *The Tipping Point*, Malcolm Gladwell outlines a theory of social movements as epidemics. Certain societal trends will roll past a tipping point, and when they do they'll explode with a momentum totally unexpected by typical economic, statistical, and societal models. Of course, we would like the global warming movement to be one of those movements that "tips" over the nebulous edge into the world of rapid societal change. But whenever discussions turn to how to "tip" the movement, the talk seems to get everyone mired in looking for the silver bullet for our cause. We'd like to suggest a different strategy—jump in and get going, and forget about the hows and whens of your particular tipping point. The only way to start a movement is by *moving*.

Which doesn't mean you shouldn't take full advantage of other trends that have already reached their own tipping points. Cultural fashions can be incredibly useful tools for activist work. The first and best people to ask are young people: your kids, the high school, the college nearby. Young people are tapped in to novelty in a profound way.

The do-it-yourself (DIY) movement has been growing in the past decade, with DIY crafts such as handmade, silk-screened, or iron-on T-shirts one of the most popular ways for people to express themselves. T-shirts are also a great, sustainable way to get your message across—and they are more likely to get repeated wearings (and viewings) the hipper they are. For the quirky annual Coney Island Mermaid Day Parade in Brooklyn, New York, one activist friend of ours made iron-on mermaids for people to add to T-shirts with the slogan, "If we don't reduce carbon 80% by 2050, only mermaids will be able to live on Coney Island." Iron-on letters are available in most crafts and sewing stores; iron-on paper for inkjet printers is available from many office supply stores; and silk-screening kits are available in craft and art stores. It's easy to print out an event logo or design from your computer for your iron-on or screen, and if you want to work from a photograph you can find directions on numerous crafting Web sites. If you have a lot of people coming to your event who want T-shirts but don't have time to make their own, sites like CafePress.com allow you to upload your art so that people can order T-shirts (or coffee mugs, buttons, and more) on demand.

Young people also know the newest tools that everyone is using to connect with one another: online Web logs, Live-Journal, Facebook, MySpace, CraigsList, Google Maps, and YouTube—an entire list of Internet-based trends. We'll talk more about using Web tools in chapter 6, but the point is that you should be up on all the trends, or have others around who can help keep you hip to them. A group of students from Portland, Oregon, organized a trip up Mount Hood for Step It Up. They filmed a beautiful seven-minute documentary of

A young woman carries an impassioned, homemade sign at a march through Lawrence, Kansas. (PHOTOGRAPH BY BEN BUSCH)

their trek, added some well-researched facts about the effects on the local ecosystem and economy, and stuck it up on YouTube. Within a little over a month, their film had been viewed by thousands of people.

Work of that kind not only spread the word in the community that watches online videos, it also made the whole thing much sexier for the regular media. So keep peppering the guy from the newspaper or the TV station with your videos, your blog posts, anything to show that you're not stuck in the sixties. We wouldn't mind being stuck in the sixties—it looks like it was kind of cool. But since we're here and now, we might as well tap in to the trends we've got.

GO VIRAL

When you begin to succeed in attracting people, those people have the potential to generate their own momentum for your project, inspire more people to do the same, and build incredible momentum that you actually have very little to do with. In chapter 6, "Make It Wired," we focus on the technical aspects of going viral online. The human aspect is just as exciting and important. Momentum turns a campaign from a tedious slog into something fun to organize and fun to join. And momentum is almost always a function of speed—it can't last forever.

Consider one of the largest Step It Up rallies, held in Minnesota's Twin Cities. Organizer Margaret Levin and the local chapter of the Sierra Club got everything rolling, and to give you an idea of the momentum they generated, their list of speakers included the mayor of Minneapolis, the executive director of the national Sierra Club, two members of Congress, a U.S. senator, two high-ranking labor leaders, and a polar explorer. But it wasn't landing the big names that kept the organizers energized and ambitious in the weeks before the event—instead, it was the calls that came in from friends who didn't know that Margaret was organizing it. People would call her and say, "Hey, there's this incredible day of action called Step It Up. You need to get on board," or "There's a rally at the statehouse that you guys should really be a part of." Getting calls from people who "didn't hear it from us" is incredibly motivating and powerful because it marks the moment when things have grown beyond your personal network and gone viral. "It became organic," according to Margaret.

To increase the odds of your action going viral, however, you're going to have to *work it*.

GIVE IT AWAY

Structure your organizing team in a way that includes as many people as possible. Worry as little as possible about control and as much as possible about seeding the ground for future organizers. Let anyone who shows an interest be a co-organizer and a colleague—putting trust in others makes for stronger long-term movements. This style of leadership inevitably leads to a more ad hoc group in the moment, but we have seen this become an asset rather than a burden—especially if and when the demands on people at home and work change over time.

We've spoken about Middlebury's Sunday Night Group a little bit in this chapter. Though it may not be the model for everyone, it has been successful over several years. One central aspect in its success is its underlying structure. It's not the most complex thing in the world—in fact, its lack of complexity might be one of its greatest strengths. There's no president, no secretary or treasurer, no VP of communications. The group is continually ad hoc, and that helps it remain fresh. To the degree that leaders emerge, it is a result of natural leadership associated with a person's skills or talents around a project, or the time he or she has at the moment, rather than titles or positions. And one of the most beautiful things about the structure is the open space made available so that people who feel compelled to act can. Got an idea? Come to the meeting, hang out, have a beer and some cookies, and tell us all about it. The group works under the general banner that climate change is a defining issue of our time and needs to be addressed. Beyond that simple message, the meetings are a forum for people to run campaigns, enlist ideas, help out, and—by far the most important—*get stuff done*.

TAP IN TO EXISTING NETWORKS

You may be working ad hoc, but it's worth trying to see if you can hook up with people who have more resources and organization, and convince them, in effect, to lend you their organization for a few weeks. Here's a place that speed *really* helps—there's a much greater chance that existing organizations will detail you some help for a month than for a year. Maybe they'll be willing to let you raise money under their tax-exempt 501(c)(3) license or lend you their mailing list. A month is a short-term exciting "project," a year is a "commitment." And since you're probably not trying to start a group of your own in a month, they're less likely to view you as a potential competitor. Indeed, make clear that you're an enthusiastic amateur. Dignified groveling doesn't hurt.

TAP POWERFUL PEOPLE

Three weeks prior to the day of action on April 14, Salt Lake City's engaging mayor Rocky Anderson went to a climate change conference in Washington DC. There he met Bill, who was speaking to local Step It Up organizers. Rocky came back home amped, ready to make Salt Lake one of the biggest and best Step It Up rallies in the country.

Rocky's first goal—rope in a top-name band. So the city's community events director, Talitha Day, approached the band Los Lobos, which drew a crowd of five thousand. City environmental director Jordan Gates opened the event to over thirty vendors from local not-for-profits and environmentally related businesses. The mayor set up a cybercafe outside the venue so that event-goers could instantaneously send e-mails to Congress and letters to the editor. And because the city organized the event, permits were a breeze.

Talitha and Jordan were bowled over by how people really listened. This wasn't the half-unconscious and slightly apathetic music festival of years past. The vendors who attended said that they fielded the most informed, intelligent questions and discussions they'd heard at an environmental event. And when the crew came through afterward to clean up (here Talitha and Jordan's civic pride shone through bright and clear) there was no trash. Five thousand people and no trash. Not your average concert indeed.

Think hard about people in different positions that can help you quickly cut through red tape and other delays, whether in government, business, or not-for-profit work. And let them know about Salt Lake City—city leaders love to host successful civic events.

CUTTING RED TAPE

Need some bureaucratic muscle to get what you need? Think back to who's already on board in your area (or search your local papers' Web sites or visit Google). A state representative? A city councillor who owns a local business? These people are tapped in to local politics and can help you out. Call them or shoot them an e-mail—they're usually eager to tap in to community enthusiasms.

DON'T SPEND TOO MUCH TIME IN MEETINGS

Organizing is partly about being organized, so there's a place for meetings—mostly to make sure everyone is on the same page and knows what needs to be done next. But organizing is much more about convincing others to take part—it's about e-mailing and phoning and working the press. So meet when you have to, and beyond that leave as much bureaucracy as possible at the door.

Step It Up organizer Donna Kaminski of Warwick, New York, assembled a core crew of four organizers through e-mail and phone. They managed to pull together environmental organizations, speakers, music, art, dance, drama, and poetry, as well as the physical logistics of a rally. They even got the Girl Scouts involved. With all this activity, not to mention daily life, Donna said that the biggest challenge was getting everyone together in a room. After trying a few times to have regular meetings, they decided to communicate through group e-mails and ended up having only two actual meetings. But that, Donna says, allowed them more time to plan for the event. And they were just as able to come together virtually around their common cause: "What moved me most was everyone's willingness to donate time, effort, skills, materials, and support," Donna remembered. "In return, we all experienced a new depth of relationship with people we thought we already knew, but knew now in another way. . . . We were being carried by the energy of this movement."

Keep in mind that Donna and her co-organizers already knew one another—they were friends. So they had a good sense of what was behind a person's e-mail tone and communication (and even a slow response time), though they met face-to-face only twice. Banter, humor, and warmth are all more difficult to convey through the computer if you have never met the person on the other end. And sarcasm, frustration, and fatigue are easier to misinterpret.

Donna's group was also working toward something concrete. They weren't having a series of online discussions to devise a strategy for the next big thing—that's best done together over coffee, tea, or beer. Instead, their e-mails were about detailing tasks and responsibilities for a preplanned objective. They used e-mail to keep one another on board, in the loop, and accountable while they worked—and they avoided the lurching, halting gait that can hobble a campaign when everyone is waiting for next Wednesday's meeting to actually make decisions.

AVOID PERFECTION

Organizers often think they have to have everything mapped out in order to run a campaign or start an organization; they worry that no one will join them unless they have thought everything through in advance. That's wrong. People like to help out, they like to be involved in the decision-making process, and they like it when someone needs their input. It turns out that it's more important to commit—to jump in and say, "I'm going to organize a demonstration outside the city council meeting next month." You can put together the final pieces of the plan with other people as you go along. If that means you don't have an absolutely perfectly polished event, so be it—we're not talking about your wedding here. We're talking about an *event*, one of many actions in a movement that may play out over years. You may be involved in some or all of the campaign, or this may be your one moment of maximum participation, but don't put it off because you're worried you don't have a complete plan. Once you start, things will begin to fall into place.

If that makes you a little uncomfortable, here's a story to give you heart. Eighty-three-year-old Dorothy Mae of Ellensburg, Washington, was forwarded an e-mail about Step It Up from a friend ten days before the day of action, and she went to the Web site and typed in some event ideas, thinking we'd contact her about whether it made sense to hold an event. Imagine her surprise when she checked back the next day and her event information was posted under "Step It Up Ellensburg" on the Step It Up map. That was April 5. If there was

going to be an event with her name attached, Dorothy decided, she had better get going and organize it. In the next
week, she pulled together an event space, a banner, a green
business speaker (her grandson), a public address system, and
a crowd (thanks to some friends at Central Washington University).

Julia Smythe, who put together an action organized in
Carbondale, Illinois, offered this inspirational summary of
just jumping in and finding your feet after the fact.

> When I first decided to start an action in support of the
> Step It Up campaign, I wasn't sure how I was going to do it.
> The first thing I did was go to City Hall and fill out an ap-
> plication and pay a small fee to rent the town square pavil-
> ion, a traffic-y spot in town, for April 14. The single most
> important factor in making my event happen was network-
> ing. I simply started talking to colleagues, coworkers, friends,
> and community members. These conversations always led
> to names of people I should contact. People would say to
> me, "You should contact [so and so]," or "I have this friend
> who can help." My friends started contacting their own
> friends, and before I knew it, an irreversible movement had
> started.

You don't need to be able to see the finish line to know
that your job is to run fast. When we launched Step It Up, we
had only a dim idea of what we would actually do to bring all
these rallies across the country together. We just knew that
photos would probably be an easy option. It wasn't until

about three weeks before the big day that we actually managed to pull together a coherent plan for posting and sharing the thousands of photos that would stream into the Web site on April 14. But we proceeded with the confidence that we would eventually figure out the best course, or at least a perfectly good one.

In the end, this strategy actually leads to less work, not more. If you plan an action a year in advance, you will likely do a year's worth of work on it. If you plan that action three months in advance, you will have only three months of work, and you'll probably get almost as much success—a 10 percent larger crowd isn't worth the extra nine months of work, especially since you can use that time to organize two or three more actions. Colleen Blacklock, a Step It Up organizer from Oneonta, New York, said that her two-month effort was just the right amount of time—"Any longer might have become tedious"—and she pulled off an incredibly strong community event, including an official proclamation from their mayor declaring April 14 as "Step It Up Oneonta Day."

BE NIMBLE

Speed implies nimbleness—the ability to turn fast when events demand or allow it. When we announced "80% by 2050" in January 2007, it was a pretty radical call to action. But times change fast—in quick succession our organizing collided with the new IPCC scientific report and a crucial Supreme Court ruling and the political landscape changed. In mid-March 2007, Al Gore testified before Congress and called for representatives to cut carbon *90 percent* by 2050. Far be it from us to hold back Congress from cutting carbon more than 80 percent—by that night, Bill had a blog post up on our site stating that Step It Up officially endorsed carbon reductions of 80 percent *or more.* Two words, one blog post, and we were able to stay at the lead of the issue and tweak a central policy of our campaign within two hours of Gore's exit from the hearing room.

What's true at the national level is at least as true locally. Bay Roberts helped lead the Step It Up charge in Boulder, Colorado. As she and co-organizer Scott Reuman worked, they realized they had more people coming than their permits could handle. Bay reworked their permits multiple times as their event evolved. Once she had figured out they needed more parking, she approached the nearby Home Depot for extra space. They said sure—and a Home Depot rep even ended up handing out compact fluorescent lightbulbs at the rally.

Because the Internet moves so quickly, e-mail and the Web are the right medium for organizing a nimble event, as we discuss in chapter 6. They're not a substitute for face-to-face community, though, and so you'll have to work hard to simulate the benefits of being in a space together. More than anything else, this means *responding to queries as fast as you can,* so that you keep building the momentum and moving your colleagues on to the next task. Imagine sitting in a meeting, asking a question, and then waiting two days for an answer. Frustrating? Of course. People have different expectations from e-mail, but you have only a short window of time before their attention moves somewhere else. We made a point during the Step It Up campaign of trying to respond to all e-mails we received within a day. And when people asked for certain tools on the Web site, we put them up as soon as possible. In one day, for instance, we received e-mails from three different organizers asking for educational materials on global warming. We got in touch that day with a not-for-profit called Topics Education, and a few days later Topics' global warming materials were up and available to all Step It Up organizers.

JUST SAY YES

It's good to note that originally we didn't even want educational materials on the Step It Up site. Our sense was that many of the materials people were asking for already existed somewhere on the Web. Others were already doing that work, and doing it well. But we soon developed the motto "just say yes." Why? It takes a lot less time to say yes than to say no. Instead of being stuck justifying your decision and wasting time and momentum in recriminations, give in and move on. It's easier to quickly put materials (or links) on your site than to direct every person who e-mails you to another Web site or organization.

MAKE IT CHEAP

Inevitably, at some point in your organizing, you'll be asking yourself the following question: "Who's going to pay for all this?" For Step It Up, keeping it cheap was one of our priorities. First, because we had to—as a completely new ad hoc group, we began with zero cash in our coffers. We also knew an important thing about money, though, which is that it often slows you down. In our experience, every dollar an organization raises is often another minute or hour fighting over how to spend it. Dollars, like all resources, are best used in moderation.

Still, you're going to have to cover your costs.

1. Hold down expenses. You won't need to worry about money if you don't need to spend it. Indeed, money is often used as a substitute for creative thinking. One of our favorite stories comes from Chris Dudley in Maryland. Chris and his

Congressman Steny Hoyer (second from right) with the Step It Up activists who paddled up the Patuxent River to deliver their message to him at his home. (PHOTOGRAPH BY CHRIS DUDLEY)

fellow organizers wanted to make sure that their congressional representative Steny Hoyer, one of the most powerful Democrats in Washington, heard their message on April 14. Instead of planning a big, flashy event and hoping Representative Hoyer would come to them, Chris and his team decided to see if they could visit Hoyer. After a few phone calls, Chris got permission to visit Hoyer's home on the Patuxent River. On the big day, the group paddled a canoe right up to Hoyer's lawn, jumped out, and snapped a picture with their Step It Up banner. "The basic thing we were trying to get across," explained Chris, "is that the consequences of sea level rise in Maryland are demonstrated by where he lives, because that will all be underwater." Chris's only expenses: the materials to make his banner and a lot of hard paddling. In return, he got one of the country's most important politicians on the record supporting Step It Up's goal, "80% by 2050." That's what economists call a high rate of return.

2. Use your local resources. Of course, some events require more than a banner and a canoe. When planning larger events, costs add up quickly, but only if you're going it alone. When it comes to organizing a big event in your community, the best way to keep costs down is something you learned in kindergarten: sharing is caring. As you plan an action or campaign, think about the people you know who may have the supplies or skills you would normally pay for. Need a banner? Try calling a friend who is an artist or seeing if kids at a local school will make one as part of an art project. Don't have the money to pay for speakers and a microphone? Maybe those kids down the street with the

garage band will lend you their equipment if you let them play a song or two. Need food? See if a local business will sponsor your event in return for some free publicity.

All of these options have one main thing in common: making friends. In Ann Arbor, Michigan, Eric Stone worked with a group to help them partner with the University of Michigan. "The university was a great resource," he said. "But having a few of us who were part of the university and understood how to navigate it was helpful. If it wasn't for that, it would have been more difficult." As many scholars have pointed out, it's no coincidence that as we spend more of our own money, we tend to forget about how much our community can provide for us. By learning how to effectively use your community resources, you'll not only save money but also strengthen your group.

3. Fund-raise effectively. When it comes down to it, sometimes you're simply going to need to shell out some dough. If you have exhausted all your other options and know that you're going to need to spend some money, then it's time to do some quick fund-raising. Our friend Mike Tidwell has raised tens of thousands of dollars for his organization, the Chesapeake Climate Action Network, by hosting an annual, one-day "Polar Bear Plunge." Turns out there are plenty of people who are willing to pony up a few dollars to see their neighbors jump into the Chesapeake Bay in January. Think about fun and creative ways to get people to pledge money—especially ways that somehow connect with your community or cause.

Ideally, every member of your group or community will contribute what he or she can. For some, that means hours

Student organizations and university departments at the University of Michigan in Ann Arbor helped smooth the way for the Step It Up event. (PHOTOGRAPH BY ERIC A. STONE)

of hard work; for others, it means financial resources. When people give money, don't just take it and run. Keep in touch with them, share your future ideas, listen to theirs, and involve them in your organization just as you would involve someone who is volunteering to pass out flyers in the park. And most important, thank them for their support!

GO HOMEMADE—IT'S BETTER

If you had all the money in the world, and all the time in the world, you could be completely polished—but that can detract as much as it can add to the success of your event. Speed requires that you throw a potluck, not a formal dinner party, and if you're like most of us, an informal potluck will be more fun and more accessible than a perfectly prepared for-mal banquet—leave those to the corporations. In general, politicians and the media pay more attention to work if it seems to come genuinely from the grass roots than from some overorganized effort—just ask a congressional staffer about the impact of a handwritten letter versus a printed form postcard.

When we say potluck, by the way, we're not just talking metaphors. Lani Gideon helped organize a Step It Up action for Hampshire College students. When food vendors proved difficult to pin down, Lani and her friends decided that it would be better to save everyone the money and cook themselves. They brewed up four huge pots of lentil soup and served it with bread, feeding a crowd of more than a hundred people for around forty bucks. Printing T-shirts also proved prohibitively expensive, so Lani hunted down a friend with a screen-printing business in town and found a hundred old shirts at a thrift store. People snatched up the free shirts and were begging for the recipe to their lentil soup. Homemade and community built entices people far more than corporate gloss ever can.

3

MAKE IT COLLABORATIVE

Our tendency when we organize anything is to work with the friends we already have, in the networks we already know about. And that's a very good idea—it's how everyone starts. But you can make whatever action or campaign you're planning far more effective if you work hard to reach out to people you don't already know. This sounds like common sense, but it doesn't happen often enough. We know it can work because the design of Step It Up produced dozens of fruitful examples.

Over the first few months of 2007, people would visit the Step It Up Web site to register an action—maybe they'd sign up to organize a rally in, say, Tuscaloosa, Alabama. Then, back at headquarters, we would notice that someone else had also signed up to host an action there. So we'd suggest that they combine efforts—and in almost every case they did. It turned out that we weren't just brokering partnerships but starting friendships, cross-fertilizing different pools of volunteers and resources, sometimes even combining very different worldviews.

It wasn't always easy. Lauren Johnson, a student at Siena College near Albany, New York, heard about Step It Up and brought it up with an economics professor who was also the campus director of environmental studies. He encouraged Lauren and referred her to other campus clubs for help. She talked to the college president, who responded positively and suggested she contact the campus's food services, which donated all sorts of snacks for her rally. By the end of the first week, she had also booked a speaker and a campus band.

And then, one week after getting all this done, she found out a group of students was also planning something for Step It Up 2007. This came as a surprise to Lauren, and initially she felt hesitant to approach them. She had become very attached to her idea, and she worried about having to collaborate and give up some of her plans. Nevertheless, she attended their next meeting and told the group what she had organized. The other students were impressed. They wanted to do an outdoor event on the campus quad involving giant melting ice blocks. Cool, in all ways. Lauren's concerns about working together melted, and the other group accepted her right away.

Or consider what happened with Virginia Nugent and Jill Sugrue. These two women in Vancouver, Washington, formed a veritable "perfect marriage" in the course of their planning. Virginia and Jill can't remember who called whom first, but they complemented each other well. Jill is a member of Al Gore's Climate Project, a nationwide network of lay climate experts, and she was interested in participating in Step It Up but didn't have time to manage the organizing herself. Virginia, on the other hand, had the time to help organize the rally but felt that she lacked the necessary experience. They agreed to work together, but it was still tough. At one point, Virginia called Jill and declared, "I can't do this; I'm going to quit." She even called the Step It Up office and said the same thing to us—she wanted a clean break. Fortunately, we had a good friend in Vancouver, Teague Douglas, and we put Virginia in touch with her. Around the same time, a local musician contacted Virginia through the Step It Up site and offered to play at her rally. With all this help pouring in, she sort of had to push ahead. And it turned out great, attracting more than a hundred people!

LOOK FOR SHARED PASSIONS

When we need help with something, we turn first to our friends, then to a wider circle of people who are still within our comfort zone. Sometimes they are friends of friends, but sometimes they're the people who share our faith, our associations, our passions. Indeed, it's rare to take a leap and invite people who don't share something with us, and with organizing it's usually our passion for an issue. Why *would* we be more likely to invite people who oppose our ideas? Still, it's important to include people who share your vision for a better world but choose to arrive there by *different* channels. Though it's understandable to feel shy with strangers, much can be gained from reaching out. We've heard lots of stories that explain why.

Take Alan Jenkins, a Presbyterian minister in Atlanta. He heard Bill speak at a National Council of Churches Eco-Justice

Conference in New Orleans and decided he wanted to plan something in Atlanta for Step It Up. His main goal: involve a diverse group of leaders from throughout the city.

To do this, Alan sent out an e-mail to many people in the region. Some were from environmental organizations such as Southern Alliance for Clean Energy and Atlanta Beyond Oil, others were members of the local clergy, and others were students, including one at the Interdenominational Theological Center, a historically black college in Atlanta.

With his group assembled, Alan was able to reach out, through them, to even more people throughout the city. "We had this vision of really giving light and witness to the number of people across the city who are all concerned," Alan shared, "and that we could all come together and speak with one voice as a part of the greater Step It Up movement and call on Congress, but also push and challenge and provide a hopeful vision here in Atlanta and in the state of Georgia."

THE PERSONAL TOUCH

PRACTICAL TIP

When starting a new project, identify individuals from diverse local organizations that you would like on board, even if you don't know them well. Learn what makes them passionate: What other groups do they belong to? What have their recent events, speeches, or sermons discussed? What communities do they serve? Then send them personal notes or e-mails that tie their passions to the climate change movement. Be humble and don't ask for too much—chances are everyone's very busy.

Sometimes passionate collaborators come from unexpected places. A few days before Step It Up, we heard from a group of thirteen convicts on Texas's death row. Their e-mail read: "Greetings friends and comrades, it is an honor to be here today (albeit vicariously) in support of what we see as a very noble cause: The defense of our precious environment: the water, the air, the land." Though they spent twenty-three hours each day in solitary confinement, they planned to fast all day on April 14 and to meditate on the links between global warming and social justice.

Did it matter? It did, to everyone who heard about it—which we made sure was a lot of people.

ASK FOR HELP EARLY

Alan Jenkins's story illustrates that you should cast a wide net when collaborating with others. But to make that easier, it helps to ask people for their help early on. We've recommended thinking of your event as a potluck, where you set the time and place and everyone brings his or her best dish. Now it's time to take the party-planning metaphor a bit further.

Say you're throwing a theme party. If you have a stake in picking the theme—we're fond of eighties parties, actually—you'll have more fun preparing for it. You'll likely have one of the better costumes, because you've had time to think about it. You'll likely have the time to come up with suggestions to improve the party (we'd suggest a playlist with lots of Supertramp and explode with excitement when a song comes on). The folks who hear about the party at the last minute may not even know what the theme is, and might show up without a costume, feeling left out. Or they'll have only a few minutes to pull something together and end up with a seventies costume that they grabbed out of the closet. They'll also end up wanting to host their own party the next time, rather than planning another great party with you.

Whenever you ask people to participate, it holds more meaning when they have some say in how the event unfolds. The earlier you pull people into the decision making, the more likely it is that they will become a new ally to work with in the future—even if your event is planned on a snappy time line.

ASSIGN HOMEWORK

Homework doesn't sound like much fun, especially to anyone who has just graduated from college. But here's the idea: at the end of a meeting or a round of e-mails, everyone (absolutely everyone) should have something to do. Whatever it is, by the time the next meeting rolls around, that person is accountable for realizing a piece of the group's goal. Make sure everyone has a range of possible tasks to choose from.

Why does this help build collaborations? Because each member feels ownership of the project as a whole and can feel proud of making a contribution to the group. Homework doesn't have to take the effort of a term paper— an assignment can be as small as calling one more person and inviting him or her to the next meeting or as big as obtaining funding for the action.

PRACTICAL TIP

THINK LIKE A FELLOWSHIP

Diverse collaborations work better as a loose group than as a hierarchy of leaders. If nothing else, this defuses any issues about whether one organization is more in control of an action than another, since at least some of the people involved will probably represent some sort of institution in your community.

When inviting people to participate, ask them to be your fellow organizers, not members of *your* group. Share responsibility. You'll also be more likely to pull in people who will pitch in because they want to rather than because you asked them for help.

The Step It Up Seattle team offers a good example of how powerful this collaborative process can be. In a city where many people are environmentally conscious and quite a few are full-time activists, things can get pretty intimidating for newcomers. According to Madeline Ostrander, the Seattle organizers pulled together a massive rally largely thanks to abandoning centralized control. "Each of the roughly two dozen event partners had political campaigns, memberships, donors, and reputations at stake in the community," she said, "and the memory of the 1999 World Trade Organization protests—demonstrations that began peaceably but degenerated into chaos—left some of these groups feeling cautious. As private citizens joined the effort, each wanting to channel his or her enthusiasm, vision, even 'outrage' into Step It Up, it wasn't clear whether they could all comfortably share common ground." Within days, a split emerged in the planning group, as some of the local organizations started worrying about the large number of unaffiliated people who wanted to join the action—and whom they did not know. At Step It Up headquarters, e-mails were pouring in from various sides of the debate. But because "global warming is an issue unlike any other," Madeline explained, the organizers stepped back to reassess their need to keep a tight grip on the event. "A week later, volunteers and organizers gathered, resolved initial differences, and agreed: one event, one collective voice, one powerful message."

What happened next, though, was remarkable. "Our shared passion built trust and strengthened communication," she recalled. "Staff of existing organizations began sharing details of event budgets and logistics more widely with volunteers. By late March, volunteers had deepened their involvement. . . . They wrote letters to the editor and articles in major regional blogs; submitted event listings to the local paper; hit up local businesses for thousands of dollars; contacted artists, schools, religious groups, and not-for-profit organizations all over town; and pinned up hundreds of flyers in coffee shops." They ended up with one of the most diverse activist partnerships in the history of the very active Pacific Northwest. In Madeline's opinion, Step It Up offered Seattle "a chance to find common ground again and reconnect with citizen roots." Because no single organization was in charge, years of competition and secrecy got swept away—no one was going to beat the others out of members, money, or media attention. Instead, everyone was getting a movement.

In Seattle, Washington, a volunteer donned a polar bear costume to humorously—and photogenically—draw attention to the effect of global warming on the Arctic ice shelf. (PHOTOGRAPH BY RANDOLPH SILL)

PRACTICAL TIP

FIND FELLOWS

To recruit lots of people to the global warming movement, you need to set aside rivalries and forget what you don't have in common—and focus on what you do. Who should you ask?

- Environmental organizations (of course)
- Schoolteachers, especially in science and civics (teachers are a great way to get word out to students and other young people)
- Religious groups that have dedicated themselves to public service in the community

- Outdoors clubs, from mountaineering groups to Boy Scout and Girl Scout troops
- Farmers and other local food producers
- Local restaurant owners, especially those who specialize in local foods or vegetarian menus
- Residents of neighborhoods near dirty-energy sources— they have to deal with more of the effects—and social justice groups working on their behalf

GO LOCAL

After you have your list of fellow thinkers, you may still want to look for ways to get more people involved in what you're planning. Find the other groups in your community who have regular meetings. Maybe it's an environmental group, or maybe it's a volunteering group or a humane society. The group should have a goal that doesn't conflict with what you're doing and an infrastructure with a track record of success. That way, you kill two birds with one stone—not only will you connect with people who are active in your community and have an established network in place, but you get to extend the reach of the climate change movement by contacting a group less tied to it.

As Kathy Hardiman of Olean, New York, told us: "Three of my friends and I were able to engage students from three colleges, a middle school class, a church group, a dancing group, and a hiking group." The church group probably would never have met the hiking club, and neither might have come across the political dance troupe—but they all turned out to have members interested in joining the fight against global warming. In Jackson, Michigan, organizer Pegg Clevenger noted that they built a "coalition of outdoor organizations—nonprofits with varied missions but who all depend upon a clean environment to pursue their outdoor passions." The volunteers worked independently and "were too busy to argue about general organization." Too busy is exactly where you want to be. And Carol Newcomb-Jones of Fort Myers, Florida, was struck by her fellow community members' willingness to help.

A week before my rally, I suddenly realized we would need a stage, and I sort of freaked out. But it was amazing how it all fell into place. I was downtown for the day, and it was a Saturday, but I went ahead and rang the doorbell at the Fort Myers Convention Center, and somebody came to the door (amazing for a Saturday!). I asked them if they had risers I could use, and they offered them to me at no charge. I needed help carrying them, so they offered to have a staff member meet me the following morning. I asked my husband and some of his friends to pick them up at 9 AM on Saturday, and together they built press risers and a stage. I didn't have to pay for anything—not the stage, not the sound system. We only had to rent lights and a podium.

But even that ended up coming at no cost. I had invited a representative from a local appliance store to come display their Energy Star appliances, and he arrived with $400 in his pocket and gave it to me, and it covered all the remaining costs. We pulled it off so cheaply. I only paid $400 for absolutely everything.

In Fort Myers, Floridians gathered to hear Senator John Edwards and other speakers at an event that cost only four hundred dollars to organize. Most of the sound and stage equipment was donated. (PHOTOGRAPH BY JOE KAKAREKA)

Not only that, but presidential candidate John Edwards turned up at her rally and gave a rousing speech!

LOCAL LOGISTICS

Chances are, if you're organizing an event, conference, or meeting, you'll need to obtain a venue. You may need a stage, a band, a sound system, or microphones. Because this is a fight against global warming, you should seek help locally for anything that requires transportation—whether it's getting supplies or getting an audience to your action.

PRACTICAL TIP

- Check with local churches, synagogues, mosques, and other religious meeting spaces; many offer free accommodations for community members on days when they do not hold services.

- Use the Web wisely. While you can Google search "Sound system rental+Amherst, Massachusetts," you won't be able to sort through all the results very easily. Ask friends. And turn to more specialized sites like Craigslist.com and your local Freecycle e-mail list (often hosted as a Yahoo! group based in cities or regions), which allow you to post requests as well as scan offers. It might be easier, and better long term, just to ask a friend. And there are lots of friends you don't know yet at places like Craigslist.

- Permits for public spaces take time. Contact your local Parks and Recreation Department early and ask about its policies. Sometimes permits cost money or require that a company or organization with insurance coverage be involved. You may need time to raise the permit fee or find a permit sponsor.

- Don't hesitate to make the occasional crazy query—someone might say yes!

THE SPEAKERS BUREAU

Another way to extend a collaborative fellowship is to invite people to speak at an action—particularly anyone who shares your passion or brings persuasive expertise to the global warming movement but can't spare the time for even a snappy event. Most organizers immediately come up with a short list—the local politician, the local environmental group's director, the local climate scientist. And it's true: to make your action credible, those speakers might play a central role.

But your main goal in recruiting speakers is to find people who will interest the audience you are trying to attract. If you want to draw business owners to get them to go carbon neutral, tap important clients or customers—or their kids. If you're hosting a panel, aim for a range of speakers that draws on every part of your diverse collaborative group.

Inviting politicians can be tricky, and not just because they have busy schedules. Politicians will deliver their own platform, regardless of what you want them to say. That's okay, and expected—but introduce them in such a way that they have to respond to your agenda before they move on to their speeches. If you have something you want a politician to endorse—say, "80% by 2050"—then put the pledge on a big piece of cardboard and ask the person to sign it. (And then get everyone to cheer—politicians love to be cheered.) Remember that these people work *for you*. There's no need to be offensive, but also no need to be in awe.

ENCORE! ENCORE!

To ensure that your speakers and your audience both walk away happy:

- Ask speakers for a commitment as soon as you have a date, and confirm the time and location a week or two before the event and again the day before (with detailed directions).

PRACTICAL TIP

- For a series of speakers or a panel, make sure everyone you invite will be equally engaging. As insurance, find the most engaging moderator or emcee in your group's network—perhaps a professional entertainer or seasoned interviewer who can keep the audience pumped or pull great answers out of a panelist.
- Brief is best; no one ever complained that a speech was too short. Let your speakers know up front how long you would like them to speak.
- If your chosen speakers include politicians, thank them for what they have done on climate change in your introduction, and express clear and resounding confidence that they'll keep leading the charge.

DO EASY FAVORS EASILY

Look for easy ways to help the people who help you. If you ask a local group to donate money or services to your event and they ask for something easy in return—like publicity—give it. For example, you might ask an outdoors club to help spread the word to their members or a local band to provide a sound system. They might say yes, but on the condition that you put their logo on your Web site, the kind of thing that helps them and doesn't hurt you. Say yes, and fast! In fact, don't even wait for them to ask. Offer it to them first.

During the national Step It Up campaign, we did precisely that. We assembled a list of "friends and allies" on our Web site, everything from huge national organizations to small groups of local animal lovers. Every single group was grateful to get the free plug, and in such good company, and we were grateful to them for their support. Gratitude is an inflation-proof currency—doing easy things fast is a low-risk, high-return investment for the future. One day you're trading logos, and the next you're cosponsoring a major news-making symposium.

Don't get hung up on whether you agree with absolutely everything a partner advocates or does—it's enough that you agree on the topic at hand. Remember, this is an ad hoc partnership.

Doing favors is also a good way to quell tensions. If you have created a Web page for your group or action, it may have a Web log or other venue for writing under a byline. When a person—we all know one—insists on making the same political speech at each of your logistical meetings, ask him or her to write an article about it so that everyone can have the opportunity to hear and consider the point. The chance to be heard in public will often be enough to get everyone back to work on the tasks at hand.

LOOSE COLLABORATIONS STRETCH

Putting these ideas into practice on a short-term project can have long-term effects. As Art Plewka, in Whitehall, Michigan, learned:

What was so fulfilling was to see the number of pieces that just seemed to fall in place once we got rolling. It truly felt like that "critical mass" Al Gore referred to. The gratitude that people continue to express to me is very humbling. The people I met as a result of this action are a big part of the positive side. We've begun to develop a network of the Michigan organizers and other concerned individuals to try and push our state legislature along toward a renewable energy standard.

At its heart, collaboration is about friendship. It's less about hierarchy and more about the joy of working with others. Jennifer Nu had just moved to Juneau, Alaska, when she heard about Step It Up 2007, and her involvement had the

With more than a hundred attendees from all over Juneau and southeast Alaska, a morning rally was held at Mendenhall Glacier in tribute to the world's disappearing ice wonders. (PHOTOGRAPH BY RICK TURNER)

unintended consequence of catapulting her into the local scene. Even though she was new in town, people, she said, "encouraged me to make the events bigger than what I had originally planned." Often an individual needs to get things started, but the realization that you don't always need to carry all the weight paradoxically means that there are times when you feel strong enough to really lead—to put everything you have into a campaign for a couple of months, confident that there will be someone else to take the reins in the future.

One of the things we have learned is that you can become acquainted with folks by going to parties with them or sharing a common interest. But you become *friends* with someone by working with them and *depending* on them—as you and your fellow organizers will discover.

4

MAKE IT MEANINGFUL

There's a simple reason and a complicated reason to concentrate on the moral depth of the action you're planning—to make it connect with the heart and the soul as much as with the brain.

The simple reason is, it works. To opponents of the climate change movement, power is mostly a function of money and the influence it can buy. ExxonMobil made $40 billion in *profit* last year, which guarantees it political clout. Community activists will never have that kind of cash, but if onlookers come to see our struggle as a moral witness, they will be swayed nonetheless.

That may sound manipulative—"Let's appeal to morality for strategic ends." But, in fact, it's the opposite. The complicated reason for morally meaningful action—one we've found over and over again—is that the more you organize actions with an eye to their emotional depth, the more you deepen your own commitment and perhaps even your spiritual growth. It's a good idea to march for five days because others will respect the sheer commitment—that's strategic. It's an even better idea because it will transform you a little in the process. And it will help create a purposeful and soulful community that has both the hope and the passion to continue this movement into the future. That's way beyond strategic.

Climate change, at some level, is a problem in chemistry and physics. Not every action needs to have an explicitly moral dimension; it's good to have rallies and teach-ins that concentrate on the purely rational reasons we need change. Fighting global warming also has a built-in level of self-interest—we need to protect our homes and livelihoods from the coming ravages. Therefore not every part of a campaign needs a religious tone or moral vocabulary, though it can be extremely helpful. But the strongest parts of your campaign will likely be those that push beyond self-interest to appeal to something deeper in all of us. Just as the civil rights movement drew on our innate sense of fairness and desire for brotherhood, so must the fight over global warming. Until it's understood to involve justice for those in poverty, a future for generations yet unborn, and a commitment to the rest of creation, it's unlikely we'll be able to overcome the status quo.

So here are a few suggestions for how to think about organizing in ways that will deepen, not just broaden, your reach. Some are about what you might call the *mechanics* of injecting meaning into a movement—walking long distances, for instance. And some are about ways of thinking about global warming that go beyond the pragmatic and rational. But remember—the strategic and the transformational are inextricably bound together, so the separation between these sections is more apparent than real.

THE FIRST (MILLION) STEPS

We have already described briefly the genesis of Step It Up. It grew from a walk across our home state of Vermont, which in turn grew out of Bill's plan to go get arrested on the steps of the federal building in Burlington.

When we started planning this kind of mass pilgrimage, we weren't really thinking about why it seemed so important to us to take a walk, but in retrospect it makes real sense. For one, walking is remarkably carbon neutral, just the kind of thing we need more of. Our medium was our message. For another, walking seems to connect with the history of human desire for change, either in ourselves or in our society. Bill had just returned from a reporting trip to Tibet, where he had watched pilgrims prostrating themselves for hundreds of slow miles on the way to Lhasa, to Mount Kailas, and to India, where he kept noticing echoes of the great Gandhian pilgrimages.

We also chose to walk from a place of meaning to us. Our guru was the poet Robert Frost, who wrote his most famous poem about "The Road Not Taken." We began our walk at his summer writing cabin in the tiny Green Mountain town of Ripton. It meant a slightly harrowing first-day walk down a steep and twisty stretch of mountain road, but it was worth it for the overtones it offered. (Everyone respects great poets, even if they don't read them.) We started on the Thursday before Labor Day weekend 2006, with some three hundred Vermonters striding east and then north, full of energy and anticipation for the five-day, fifty-mile trek.

Whatever we had expected, it turned out better. New people joined us every day, a gathering throng. Friends, many of them new, generously offered their homes and fields along the route as resting places for the night, cooking delicious meals of local food—one night we ate pizza from a brand-new clay oven a wheat farmer had built by his barn. And we gathered on small-town greens each evening to share stories of our walking, as well as poetry, music, inspirational words, and hopes and dreams for political and social change.

Each stopping point became a time for celebration, partially for the break from walking mile after mile on hard concrete—it's actually tough work, and before long our feet were sore. But our spirits were high, because we sensed, as we walked, that we were having a tremendous impact. And not just on drivers passing us by, honking and waving, or the people reading the extensive daily accounts in the newspapers, or hearing the story on the radio. We also knew we were affecting *ourselves* in interesting ways. The act of walking—the physical challenge, the opportunity to see the state in which we live at a much slower pace than when zipping around in cars, and the chance to talk for hours on end with inspiring and dedicated companions—was about as fulfilling and exciting as any endeavor we had been a part of. Simply the chance to hear a stranger's whole story instead of "cutting to the chase" seemed a wonderful and transforming luxury.

The walk kept gathering power. On the final day of the journey our line was a mile long marching into Burlington, the largest political gathering about *anything* for many years in the tiny Green Mountain State. The crowd was plenty large enough to convince all the candidates for federal office in that fall's election to come meet with us, and every last one of them, even the most conservative Republicans, signed a pledge committing to support policies such as an 80 percent reduction in carbon emissions by the year 2050.

THE POWER OF CHILDREN

P R A C T I C A L   T I P

If you want a politician to sign something, get a young person involved. In our case, Schuyler Klein, the youngest person who had gone the whole distance, handed each of the candidates a marker as he or she got up to speak. In fact, involve kids in all kinds of ways—they're the ones who will know someday how well we did our job.

It was an extraordinary five days—if you'd like to see some images of it, a filmmaker named Jan Cannon has produced a very fine video (available at www.JanCannonfilms.com). But its true meaning lay in the dividends it produced. The people that Vermont voters sent to Washington, notably new senator Bernie Sanders and Congressman Peter Welch, have been on fire about global warming ever since they took office. (Sanders introduced an "80% by 2050" bill as his first piece of legislation, on the first day the Senate was in session.) And our walk has inspired others to do the same, with just the same

Religious activists joined the Interfaith Walk for Climate Rescue from Northampton to Boston, Massachusetts, to demand an 80 percent reduction in global warming pollution by 2050. (PHOTOGRAPH BY ROBERT JONAS)

kind of inspiring results, in Massachusetts, New Jersey, and Oregon.

You can read about the Massachusetts walk, the Interfaith Walk for Climate Rescue, online, too, where you'll begin to see the power that religious tones and themes can bring to organizing efforts (www.climatewalk.org/blog). The eight-day walk was organized by Religious Witness for the Earth, which has been building an interfaith environmental movement since 2001. The March 2007 walk was their latest and largest endeavor yet—and they were tested from the start, since they took their first steps from Amherst toward Boston in rare late-winter blizzard conditions. Undeterred, the walkers made their way town by town, sleeping and eating in the refuge of churches and synagogues along the route, praying, singing, and walking together. Lynn Benander, blogging part of the journey, described the feeling and progress of the group.

> We were basking in the sun with a brisk breeze at our back the whole way. . . . There's a spirit of generosity from everyone. We're very much enjoying this experience of harmony in community. (Quite miraculous actually, when you note the strenuous physical activity, total lack of personal space, sleeping bags on hard floors, aches and pains, minimal bathroom facilities and no shower facilities!)

Farther along in the trek, Rev. Margaret Bullet-Jonas made a new young friend. We'd like to share her story at length.

> A little boy walked up to me, looked into my face, and slipped his hand in mine. For most of the rest of the way into Cambridge, we walked hand in hand. He told me that he was five years old, and that his name was Alden. His mother walked nearby, and although from time to time Alden climbed into the stroller and took a ride, for much of the day we walked the route together.
>
> I don't know why Alden attached himself to me, but the affection between us was immediate and it wasn't clear who was receiving the greater gift.
>
> "He's giving me energy," I explained to a fellow walker who wondered why the boy was with me. "Energy is coming out of his hand into mine."

When we reached the bridge across the Charles River that leads into Harvard Square, Alden asked me to lift him up so that we could look over the railing. Together we studied the river below and the skyline of Cambridge and Boston beyond. When I told him that we were walking to Christ Church and that its address was Zero Garden Street, he burst out laughing.

"If the address were Zero Road Street, then there wouldn't be a road there at all," he told me, grinning mischievously. "It would only be grass!"

As we set out on the last half mile, I looked down at this quirky, adorable kid who seemed to have come out of nowhere and walked straight into my heart. I looked over at his older brother, Andrew, and I thought of the other children who had joined us at some point along the way. I thought of my five-year-old grandchildren, Noah and Grace, and I thought of all the children, born and not yet born, on whose behalf we make this walk. It is our love for these children that sent many of us out into the streets to press for immediate, stringent reductions in greenhouse gas emissions. If we can just stay connected to that love—and receive its energy into our hands and minds and hearts—who knows what we will be able to accomplish together.

No one "organizes" moments such as these. But moments, experiences, and new relationships like this one are made possible by the spirit and energy dedicated by those who organized the walk and the individuals who took the risk of leaving their ordinary life behind for a little while.

The Oregon walk was part of Step It Up 2007 and began in Portland on April 14 with a large rally at the opening of the journey rather than at the end. A few weeks afterward, when we spoke with organizer Martin Tull, who works for the Northwest Earth Institute, he still seemed to be on a high from the events. He described it as the "best rally in twenty years" and attributed much of its success to the rally's emcee, Jefferson Smith, the executive director and founding chair of the Bus Project (www.busproject.org, a creative and impressive organizing model of its own). Jefferson ran around the audience, engaging them with questions about why they cared about global warming, a tactic that proved to be far more energizing than speeches about policy and legislation.

The Oregon walk then transported that energy and carried it forth for four days and over fifty miles to Salem, the state capital. It was not a mass march, just ten people—proof that a pilgrimage does not have to be huge to be a success. Martin described it as a "personal spiritual journey" for them all, and noted three insights. First, they too experienced a vastly new perspective on "car culture," seeing roads and highways from a walker's pace. Second, they gained a dramatically new appreciation and awareness of the beautiful scenery around them as they simultaneously walked on hard concrete—a strange and transformative blend of natural beauty and intimacy with our industrial culture. And third, it was physically demanding, enough so that it felt like a real accomplishment for them all. And as they held a well-attended press conference at the statehouse and members of the state legislature used the walk to further some of their climate-related political efforts, their success and power crossed into new forms.

As we finished writing this book, college students were planning summer marches in the key primary states of New Hampshire, Iowa, and South Carolina. Our guess: there will be many more to come. Maybe you'll organize one.

SAFE JOURNEYS

It's always best to have the whole group alive at the end of any project. Dead activists aren't the story we ever want to see in the news. Safety is paramount.

- Designate specific people as the lead and the tail of your group and charge them with alerting oncoming traffic— brightly colored clothing or signs are useful—and with watching for individuals who are having difficulty keeping pace.
- Keep people out of harm's way. On roads, remain on the shoulder or the sidewalk.
- Pack first-aid supplies, recruit a health professional to attend the action "on call," and let your local authorities know your plans in advance (assuming those plans do not involve lawbreaking).
- And pack *a lot* of water—you'll need it.

PRACTICAL TIP

MOVE ANOTHER WAY

If you don't want to walk, then bike—or skateboard, or row, or anything else that helps remind people that their muscles are sufficient to get them where they need to go. We're used to the idea that we use our muscles for "exercise" and "recreation," and we use our cars for anything serious. But you can blend the two by making your action full of motion—somewhat the same way that the Critical Mass campaign has been using big group bike rides to advocate for bike-friendly cities. Or take the example of one activist who kite-boarded across North Dakota to demonstrate the abundant potential to harness wind power across the Great Plains. And don't forget about America's great waterways, which will be under threat from flooding—there are many wonderful places to canoe, kayak, or sail for change.

Twenty-three activists sponsored by Exchange Cycle Tours bicycled to a rally in downtown Portland, Oregon, stopping at the People's Co-op, a passive solar house, and two other events along the way. (PHOTOGRAPH BY DANE SPRINGMEYER)

There is something quite powerful about any kind of endurance in our culture—in an age when most people go all year without walking a mile, the mere willingness to do anything for an impressive number of miles or days or hours makes a certain statement. This counts!

OTHER WAYS TO COUNT

Sometimes it's just not possible to get people or things from one place to another without burning fossil fuels—even if you're walking one way. You can calculate how many tons of carbon your action will emit by using a carbon calculator and then purchase carbon offsets, known as Renewable Energy Credits, to compensate at a Web site like Native Energy (www.NativeEnergy.com), which invests the money in clean wind power and methane digestion projects. Being aware of your "carbon footprint" at all times is a morally meaningful project.

PRACTICAL TIP

EXPLORE SILENCE

Just as we're accustomed to constant mobility—and thus it's strange and liberating to walk for five days along a road you would otherwise cover in an hour in a car—we're accustomed to constant noise and chatter. Falling silent for a while can be quite radical, and it can change your mood dramatically.

A rally, for instance, doesn't have to be constantly noisy—sometimes it's more thoughtful when the sound dies down. At the big rally that ended the Vermont march, we were a little worried that some of the more conservative politicians might get booed for their stands on other issues, or if they didn't endorse our global warming goals. We don't shy away from confrontation, but that day it was important to us that the mood be nonpartisan and equal parts solemn and joyful. So before anyone got up to speak, we had the crowd practice being silent instead of booing in case someone said something they didn't like. They made no noise for a minute—a "crushing" silence. And the afternoon proceeded without any heckling and with much good cheer.

CELEBRATE LOCALNESS

One of the ways to add to the moral weight of your action is to imbue it with the weight of local affection and local history, as we did when we called upon Robert Frost's "Road Not Taken." Likewise, Step It Up organizers in Salt Lake City gathered waters from all of the streams that surround that dry city and ceremoniously mixed them.

Place is one of the things we're losing in this country, as it homogenizes into endless subdivisions and malls. But localness has enormous attraction that crosses political lines. It's why one of the most useful symbols we have is the farmers' market; a feast of local foods helps make any action more understandable. It's about home, it's about self-sufficiency, it's about tradition—it's not about liberal or conservative. And in many cases, local food helps illustrate the point to be made about climate change. Farmers are usually the keenest observers of changes in the weather, and many are truly freaked out by the advent of climate chaos. In the Northeast, some of the most outspoken activists have been the maple syrup makers whose entire livelihood depends on the continued steady progress of the seasons.

Talking about a sense of place can resonate in unexpected ways. In Boston, Step It Up organizer Brooke Muggia went to speak about climate change to a group of kids in one of the city's poorer neighborhoods. She reflected, "Their experience is very different than mine. I've had the opportunity to travel around the world, and these kids are growing up with cement and concrete in the projects, broken glass and brick walls to shield them from gunfire." Unsure how to start her talk, Brooke asked the students to tell the class about a place that was special to them. "The hands that went up! We had a conversation for two hours," she recalls. One child in particular struck her.

> At one point, a young kid raised his hand and said, "The laundry." And I said, "What do you mean?" All eighty kids in the classroom began to laugh. But the boy continued, "When I go to the laundry mat with my grandmother, I close my eyes and hear the waves of the ocean. And when I open my eyes I can see the waves crashing and the sharks swimming among them. This is where I go to experience the outdoors." The boy came up to me afterward and said, "Thank you for not laughing."

List the things that make your place special, and then think about how some of those deeply resonant traits might

To emphasize the need for carbon-neutral electricity, dozens of canoes and kayaks paddled on Mill Pond in New London, Minnesota, the "land of 10,000 lakes." (PHOTOGRAPH BY JEFF VETSCH)

be brought to bear in your organizing. We can't stress this part enough. For reasons both tactical and meaningful, a sense of place is invaluable for effective organizing and for creating a brighter future.

CHANGE VALUES

Here we switch—sort of—from tactics to ideas (always remembering that tactics *are* ideas). Talking about values is often a touchy subject, especially recently. In the last few years, "moral values" have entered the political arena, and pundits, politicians, and pollsters have done their best to dissect them and put them in boxes. You are a conservative; your neighbor is liberal. Your mother believes in an almighty God; your father is a heathen. Our values have been so sliced and diced it's hard to see all that we have in common.

We see the fight against global warming as a way to bridge the ever-widening gaps in our country and help create common ground. We're never all going to agree about everything, but we may be able to agree that stopping the planet from going kaput is a good idea. Just about the only *good* thing about the global warming crisis is that everyone has something at stake. Certainly, some have much more at stake than others, but a diversity of people, and a diversity of values, are uniting to take on climate change.

Here are a few ways to talk about climate change as an issue of values—some moral, some political, but all powerful.

CARING FOR OUR CHILDREN. Only one of us writing this book has kids (yet), but all of us are worried about the world that future generations will inherit—that's the most natural of human feelings. Many people are motivated to fight global warming by the feeling that to do otherwise would be an injustice to their children or grandchildren. Quite a few of our Step It Up events involved kids—as speakers, performers, or representatives of the generations to come. Just like many smokers will quit before having a child, many of us will fight our addiction to fossil fuels because we have the next generation in mind.

HEALTHY COMMUNITIES. Much has been written about the loss of community in America. As the sociologist Robert Putnam pointed out, we are increasingly finding ourselves "bowling alone." Thankfully, the solutions to global warming are often also the solutions to fractured communities, whether that means farmers' markets or mass transit. Talk to people about what ails your community and see if you can work for solutions that help address that problem as well as global warming. This might require some creativity, but it often gets many more people involved.

ECONOMIC OPPORTUNITIES. There is a tremendous amount of excitement right now about the potential of a new clean-energy economy. We think it's warranted. Author and activist Paul Hawken says that this generation has the opportunity to completely redo the way we do just about everything. That process is going to create a lot of jobs and stimulate economic growth. Ending our addiction to dirty energy won't harm the economy; it will help it. And it may help most the people who need help most—one of the best plans we've heard about is for a Green Jobs Corps that would train unemployed youth to do things like install solar panels and insulate homes.

And this is just a start. We encourage you to think about what people in your community value and how these values might inspire them to join you in taking on global warming. Remember, your neighbors might not have the same reasons to care about the issue as you do. That's okay—and frankly it's a good thing. Having a diverse group means you'll be able to reach out to different people more effectively. Agreeing on why climate change is a problem is less important than agreeing on the need to find solutions.

A MATTER OF FAITH

For many people, values take the form of explicitly religious injunctions and images, while others may be uncomfortable with any talk about religion. Regardless of what camp you might fall in—and we believe climate change is an issue that can help unite the varying camps—the rising involvement of religious leaders and institutions in the political movement to fight global warming is an important shift. As many as 85 percent of Americans identify themselves as Christians, and many of the rest are Jews or Muslims or adherents of another faith. The story of their increasing environmental involvement is both fascinating and hopeful.

For a long time, no religious environmental movement existed in this country. Liberal churchgoers tended to view the environment as a luxury to be addressed once we had taken care of war and poverty; conservatives tended to think that anything green was altogether too close to paganism for their comfort. There was Saint Francis blessing the animals, but that was about it.

Change came slowly, and with much effort from many people. Evangelical leaders like Calvin DeWitt slowly developed a theology of creation care that led conservative Christians to the biblical messages on stewardship. Academics like Mary Evelyn Tucker spearheaded conferences at Harvard that gathered researchers from each of the world's great religions to plumb their scriptural and traditional resources. Paul Gorman and the National Religious Partnership for the Environment worked with leaders in Protestant, Catholic, and Jewish denominations on everything from policy to liturgy. The remarkable Interfaith Power and Light group began persuading congregations to cut their energy use. And some of this work slowly became more overtly political. In 2000, Bill helped organize a demonstration of clergy outside Boston SUV dealerships with a big banner asking "What Would Jesus Drive?" Within a few months, the Evangelical Environment Network had picked up the slogan and was running television ads in many states—ads that scared Detroit and jump-started a long process of deglamorizing the megacar.

It wasn't until the winter of 2005, however, that things shifted into high gear. A group of leading evangelicals, including the presidents of many seminaries as well as pastors such as Rick Warren, signed on to a statement declaring that global warming is a real and serious problem. Rich Cizik, chief lobbyist for the National Association of Evangelicals, said it was an issue on which his group would pressure Congress and the White House. This was not welcome news to all in the fold. Some of the most prominent televangelists preached against it—Jerry Falwell, a few months before his death, said global warming activism was a trick of Satan designed to move the focus away from issues like gay marriage. But most of the evangelical leaders have stuck to their position, and indeed become more forceful. The power of this witness should not be underestimated—this was the first issue over which they had broken with the Bush administration and the general Republican orthodoxy. And it offers a chance for those from other backgrounds to build some bridges. You can focus on either what you disagree on (creationism versus evolution, say) or what you agree on—that it would be wrong to de-create Earth, no matter where life came from.

In any event, the landscape is changing—organized religion in America now confronts the issue of global warming squarely. COEJL, the Committee on Environment and Jewish Life, has done great work in its community; most of the mainline Protestant denominations have taken strong stands; in the Roman Catholic community, important leadership has come from, among others, convents and monasteries. In the summer of 2007, several religious organizations, including the National Council of Churches, the Islamic Society of North America, the political arm of the Reform branch of Judaism, and Episcopal, Methodist, and Baptist churches, released a statement calling for real action on global warming from the federal government.

In local communities, many preachers have started making these issues a focus of weekly worship. When we were organizing Step It Up, we received help from leaders ranging from Jim Wallis and his Sojourners Community to a Connecticut boy who decided to turn his bar mitzvah celebration into a global warming rally. *Christian Century*, the chief magazine of the mainline denominations, devoted a cover story to Step It Up, and its editor wrote a special appeal to churches asking that they organize actions on April 14. That did the trick: on April 14, there was an extraordinary turnout among faith communities across the country. Teach-ins at churches in Hawaii and Florida. Interfaith coalitions offering compact fluorescent lightbulbs to their communities or delivering letters to politicians. Earth Keepers, in Marquette, Michigan, mobilizing their network of 130 churches from nine different faiths for an action in the Upper Peninsula of Michigan.

Rich scriptural traditions in all the faiths point people in the right directions. We're not theologians, so we won't go into specifics. Instead, we'd like to refer you to Web sites such as Web of Creation (www.webofcreation.org), the Evangelical Environmental Network (www.creationcare.org), the Islamic Foundation for Ecology and Environment Sciences (www.ifees.org), and the interfaith Protecting Creation (www.protectingcreation.org) for more information. (The Resources section of the Step It Up Web site lists lots of other places to look online regarding religion, spirituality, and global warming.)

Three overarching arguments constitute the moral case for doing something about the climate. Too often the rhetoric is about what's going to happen to *us,* how *our* economy will be harmed. The moral case asks us to think about someone besides ourselves.

THE POOR AND THE POWERLESS. One of the hideous ironies of climate change is that it will strike the people first who can deal with it least, and who have done little to cause it. Those who live on the coastal plains of Asia, or in the desertifying interiors of Africa, or anyplace where mosquitoes spread malaria are on the front lines—and they are universally people who emit a tiny percentage of the world's carbon dioxide totals. To American Christians, for instance, who are insistently called to "love their neighbors," this should be alarming; our luxury is another's demise. It's perhaps the biggest reason religious communities have become so involved in this fight.

THE REST OF CREATION. The guess is that global warming may put a third or more of all the plants and animals on Earth in danger of extinction—if it goes on long enough, it may be as thoroughly destructive as the last major asteroid to crash into the planet. For a long time, people have struggled to explain why extinction is bad. Secular arguments often seem a bit strained—we need to save species because something in their DNA might someday cure some illness you might someday catch. But consider the religious argument made in a series of TV commercials from the Evangelical Environment Network: God made it, so who are we to wipe it out? Given that the first commands in Genesis are to take care of the Earth, dressing it and keeping it and exercising benevolent dominion, our current spasm of reckless behavior seems unbiblical in the extreme. Noah emerges as the first "green"— saving a breeding pair of everything from the ravages of extreme, if temporary, climate change.

THE FUTURE. If we owe a moral debt to anyone, it's probably to those who will inhabit the planet in the future. A philosopher might call this "intergenerational equity," and an economist might attempt to dismiss it with talk of varying the discount rate, but most people feel a deeper moral call—

given what past generations have sacrificed to make our lives better, it seems indecent that we're steadily degrading the world. There are of course Christians who think that the world is about to end, and some of them are therefore skeptical of the need to protect the environment. But even in that group, many who believe Judgment Day approaches want to be able to make a good account of their actions on this lovely planet, where they believe that every tree and animal and baby speaks somehow of the divine.

We've noticed that for a great many people, even those who don't get to a place of worship all that often, the power of religious imagery is real and moving. One of the highlights of our Vermont march was a Sunday morning service at the Charlotte Congregational Church, where people were spilling out of every door and passionately singing hymns. Our march that afternoon had a special feeling to it—not religious, not solemn, but *connected*.

Meaning and tradition and religion and values are hard to talk about. They're bundled together in our culture in ways we can't entirely pull apart. All we know is that *everyone* wants to act from the heart as well as the head, and that doing so will add power to what they're saying—power for themselves and power for the outside world looking in. Global warming is *morally* serious. For justice of all kinds, for peace, hope, and the safety of our world—speak openly about the moral significance of climate change. Whether our values be spiritually grounded or otherwise, we all need to be moved at the deepest levels by this issue.

5

MAKE IT CREATIVE
(and Fun!)

Too often when we organize rallies or campaigns, we fall into ruts. We get a stage, we get three speakers, we get a microphone, we get a crowd, we get some petitions to sign or postcards to send to Congress. That's good—it's a great place to start. And if you have a truly great speaker, and a truly massive crowd, it might be enough. But one problem is that most of us will never manage the galvanizing rhetorical power of Martin Luther King Jr. And a bigger problem in our experience is that such a setup doesn't make full use of what most of us do have—creative minds that can figure out how to make an action or a campaign memorable.

Recognizing the creativity within ourselves and our organizing communities is just as critical as raising enough money to pull off something big. *Effective actions are supposed to make people think outside the box, and so they need to be out of the ordinary.* You can announce that you're organizing an "Earth Day rally with speakers" only a few times before it completely loses its freshness, excitement, motivational power, and fun.

Our world is changing at a breakneck pace, and as activists we need to keep developing new, innovative tactics to get out messages and flex grassroots muscle.

We were amazed by the sight of the pictures flowing into the Step It Up Web site on the night of April 14, and not simply because there were so many of them. In Claremont, California, for instance, organizers gathered materials from the recycling bins at five local colleges and as people listened to speeches, they also made bowls, lamps, and bookshelves out of trash. It wasn't just a voice intoning, "We need to live more lightly on the earth"—people were living, at least for a moment, more lightly on the earth. In Key West, Florida, organizers were worried about the effect warming waters would have on the coral reefs that fringe the Keys. They could have gotten a marine biologist to give a talk about the way increased temperatures would bleach the corals, and that would have been pretty interesting. Instead they found a bunch of scuba divers and held an underwater rally at the coral reefs themselves, and millions of people across the country who saw their video came to understand in a deep and visceral way why they were taking action.

Creativity is the low-cost, high-impact solution to getting your message out. It also brings people together by nourishing them as organizers and citizens. As the activist theater group Bread and Puppet says in its *Why Cheap Art? Manifesto,* "Art is food. You can't eat it but it feeds you."

USE MUSIC

Music is a big part of our lives, and in the past it has provided a lot of the spirit for social change—it's hard to imagine the civil rights movement without the freedom songs that helped give people courage and solidarity in the face of brutality. But environmentalism has never been a particularly musical movement; it has tended to be highly rational, to make more use of statistics than perhaps it should, and less of guitars and drum kits. That's one of the things we wanted to change with Step It Up, and climate organizers around the country succeeded in making it happen. Here are some of the ideas we've culled from their experiences.

Big concerts with varied acts are a great way to have fun, get the message across, and reach people who wouldn't otherwise come to a rally. In chapter 2, we described the Los Lobos concert in Salt Lake City that attracted more than five thousand people, but there's no need for a big-name band. Lisa Dollinger, an organizer of a solar-powered concert dubbed Solar Rock, held in Tucson, Arizona, said that a high school rock band that performed at the event attracted scores of sixteen- and seventeen-year-olds, a group she wouldn't have reached otherwise. Having local musicians perform fosters community connections that will make your rally and campaign stronger. A concert will entertain your core audience, attract passersby, and get musicians involved. Often three or four songs performed between speakers or activities are enough from any one act. And don't forget to send thank-you notes to musicians after the event—they have (we hope) given you free what they're used to being paid for.

Remember that you can also be creative about the kind of music you offer. In Providence, Rhode Island, the Brown University band led a marching column of college students to the steps of the statehouse for Step It Up. Set up a battle of the bands to battle carbon emissions, or host a "dam jam" to raise awareness about flooding.

MAKE IT AUDIBLE

For a large rally, having a stage and sound system is a must, but perhaps less necessary if you are anticipating a smaller crowd.

- Often local businesses or not-for-profit organizations will loan equipment, sometimes accompanied by a person who knows how to use it.
- In general, make sure that you've secured the equipment as far ahead of time as possible and, if you live in a potentially rainy place, get a tent or tarp to protect it.
- A 100-watt system should be enough for a crowd of up to three hundred people. If you're expecting a larger turnout, make sure you have speakers that project far enough for everyone to hear. Know which microphone goes into which input on the board (label them with masking tape) and delegate someone to adjust the volume, treble, bass, and gain throughout the event.
- Stages can be constructed out of 2 × 6 boards and plywood.

PRACTICAL TIP

- Most important, make sure to set up your stage and equipment and conduct a sound check a couple of hours before your event.

You may decide you don't need professionals at all. In Molokai, Hawaii, the "Keep Paradise Cool" rally hosted a luau that featured hula dancing and live music from local ukeleleists, guitarists, and drummers. All of the folks who participated were community members who came together to jam with friends and neighbors.

Letting kids perform conveys the message in a whole different way—suddenly, people start to realize the stakes for the future right in the gut. In Naples, Florida, which is at risk for high-intensity hurricanes and rising sea levels, a chorus of children joined local songstress Tish Poorman in performing an original song about climate change. Before long the crowd joined in. The night ended on the sand with a booming thirty-person drum circle, the final rays of the sun disappearing on the horizon. That moment, event organizer Dianne Rhodes said, was "absolutely riveting."

One of the giddiest days of the whole Step It Up 2007 campaign came when an unsolicited musical track arrived via e-mail from a band we'd never heard of called the Gallerists. They had composed a song called "Step It Up"—and it was *good*. We listened to it every morning for the rest of the spring and posted it on the Web site, too.

The Gallerists' song gave us an idea. We teamed up with project MUSE (www.musecampaign.org), an organization promoting action on climate change and other environmental issues through music and songwriting, to provide lots of other wonderful songs on the site that people could play and download for free. We decided to create a tool so that any other musicians who wanted to could upload their own songs about climate change to the site, and we streamed featured songs on our front page. That Web presence also let visitors know how much value we placed on having music at Step It Up actions around the country.

Music is important because it helps form relationships in the communities we live in. It's not enough just to tell people to cut carbon—yelling louder won't make it more effective. As the Raging Grannies, a chorus of radical elderly women in flowery dresses, always insist, we have to sing it to our families, friends, and neighbors, to our elected officials, and to complete strangers so that they will really listen to us. A singing movement, in the end, swings minds.

USE YOUR BODY

Emma Goldman, the outspoken anarchist, reputedly said, "If I can't dance, I don't want to be in your revolution." She was right: the revolution won't be much fun unless we can put our bodies into it. That doesn't mean we all need to walk fifty miles or chain ourselves to coal-fired power plants (though those aren't bad ideas). But as we saw in the previous chapter, moving around—stretching your muscles and flexing your joints—helps you stay healthy, get your message across, and is just plain fun.

In his landmark studies of flow, the great psychologist

Mihaly Csikszentmihalyi found that humans consistently report finding the most joy in two activities: dancing and volunteering. Then naturally the happiest activists should combine the two—and you'll want to be one of them. Bennett Konesni, a local musician and farmer, planned a Step It Up contra dance (New England's version of a square dance) at a high school in Belfast, Maine. More than 120 regulars, most of whom had never attended an activist gathering, and local environmentalists, many of whom had never been contra dancing, came together to do-si-do to a local string band. Not everyone knew about the Step It Up campaign, so as they twirled around the hall, Bennett passed along the message of "80% by 2050" to each of his many partners. Those people, in turn, spread the word to their partners. By the end of the night, everybody knew about the dangers of global warming to coastal Maine and the goal of passing bold and comprehensive climate legislation.

Another way to make your campaign or event fun and exciting is to do something extreme—extreme enough that it draws attention for the extreme effects of global warming. Maybe that's a hundred people on bicycles. Or maybe it's something wilder. On the April 14 day of action, Forrest McCarthy and a group of seven climbers hiked up and skied down the shrinking glaciers on Gannett Peak, Wyoming. The state's tallest mountain, Gannett provides much-needed water for the agricultural communities in the valley below and is a favorite destination for expert mountaineers to test their skills. Since the middle of the twentieth century, scientists have reported that glaciers have been receding on Gannett at an

Activists climbed Wyoming's tallest mountain, Gannett Peak, to stress that the disappearance of glaciers due to global warming will have a significant economic and social impact. (PHOTOGRAPH BY RE-BECCA HUNTINGTON)

alarming rate due to global warming. "We descended the glacier via skis to establish our last camp along the frozen Dinwoody Creek, each vowing to Step It Up, so that there will continue to be glaciers to ski down and to provide water and beauty for centuries to come," wrote Amy McCarthy, a co-organizer of the event.

Early in planning Step It Up, two of our favorite friends—the writer Janisse Ray and her husband, Raven Burchard—organized a rafting trip. It was January in Vermont, but it was 60 degrees, so they got some friends to float with them down a river that should have been frozen. "We'd rather be skiing," their sign read; friends stretched a banner along the side of a bridge asking, "Where's winter?" Needless to say, it made the papers. (You can see them in action on page 10.)

Bicycles are easier for most of us than rafts. Each year on April 1, a group of our friends hosts the Fossil Fools Bike Ride, a climate-themed cycling event. Last year, the small group rallied more than 150 people to pedal the forty miles from Burlington, Vermont's largest city, to the capital, Montpelier, in support of renewable energy legislation. Organizers handed out pinwheels representing wind turbines to the costumed and decorated cyclists as they pushed off and concluded the event with a rally on the statehouse lawn.

Some years this ride has taken place in a pouring rain—but it's always a success because by *moving* toward a clean energy future, you feel you're making a difference. A Vermont TV station, Channel 9, sent a camera crew one year, and the reporter hung out the car window for twenty miles, interviewing bikers as they pedaled. Getting attention helps you feel like you're making a difference, too.

Using your body can also show a commitment to your message, quite literally. A big part of the Step It Up actions was aerial art. It's often cheap to pull off but requires a bunch of people and a fair amount of organization ahead of time. Our friend John Quigley, a professional aerial artist, helped us put together a huge Step It Up photo event in January, where more than eight hundred students from Treasure Mountain Middle School in Park City, Utah, spelled out "Step It Up" in English and "We hear your call" in Inuit. John took a beautiful

Hawaii PV Coalition, which supports development of photo-voltaic (PV) solar energy, and the residents of Maui gathered on Baldwin Beach, one of the beaches that the island stands to lose with rising sea levels caused by climate change. (PHOTOGRAPH BY CHARLES OREVE)

photo of the students, the Wasatch Mountains nestled in the background, from the basket of a hot-air balloon. The image ended up getting published on a number of Web sites and in many places in the print media, including the *New York Times*.

Inspired by the photo, other organizers adopted the idea on a smaller scale to create equally powerful records. In Paia, Hawaii, the Hawaii Photovoltaic Coalition delineated the future tide line by standing in a line on the beach. In Clemson, South Carolina, college students artistically assembled as "CO_2" and a downward-pointing arrow while wearing T-shirts in eye-catching colors. In Washington DC, rally-goers "sent"

Congress a human postcard composed of hundreds of people in front of the Capitol spelling out "80% by 2050."

Aerial art gets people involved and spreads the message far and wide by showing that people will take time out and work together for a solution to the climate change crisis. All it takes is an outline taped on the ground; a bullhorn; a tall building, tree, or crane; a camera; and a vision.

ACTIVIST VISION

The first assumption we made with Step It Up was that images would count most. Instead of one huge rally keyed on one amazing speaker whose words would pass into history, from the start we envisioned pictures streaming in from all over the country. That's what we got, but their ability to get people involved and connected was something we had not fully imagined. It makes us think that visual art should be a part of as many activist visions as possible.

In Cambridge, Massachusetts, a group of community members led by two high school students created a stunning display to share with Congress. Dubbed "Winds for Change," their rally featured more than a thousand kite tails, designed by students at a nearby school, that communicated their wishes and dreams for a carbon-free future. The colorful tails swept from the top of a flagpole to the ground in the center of Harvard Commons. The string of kite tails had two intended meanings: a shift in our thinking on climate change and an appeal to support the clean-energy wind turbines proposed for Massachusetts' Nantucket Sound.

In Galesburg, Iowa, a group of concerned citizens constructed a giant yellow shoe to represent the volume of carbon dioxide released into the atmosphere from the burning of one gallon of gasoline. About the size of a compact car, it humorously and effectively demonstrated the carbon "footprint" concept, and evoked an image of humans trampling Earth.

In Middlebury, Vermont, a number of creative young people desperately wanted one thing: to be the first Step It Up action in the country on that climactic Saturday, April 14. One smart student came up with the idea of flashlight art. At 11:59 PM on Friday, the small group ran to the middle of a dark field and lined up in formation. At exactly midnight, they set a camera to a long exposure length and flicked on their flashlights, each spelling out a prescribed letter of "Step It Up" by moving his or her light. It took them only one try. The result was a breathtaking photo that brings to mind meteors or an extraplanetary presence. Beautiful—and you can't get much cheaper.

Art can be created from "found" objects as well—that is, things that people throw away, like the contents of those recycling bins in Claremont, California. At Middlebury's 2006 commencement, many of the same students who created the flashlight art were upset with the choice of a former Monsanto executive as the ceremony's main speaker. Instead of booing and hissing during the speech, they decided to show their support for wind power in Vermont by populating the campus's main quad with wind turbine art. They went through all the detritus of finals week, using hundreds of

Demonstrating that creative action doesn't have to be expensive, thirty Middlebury College students took a long-exposure photograph while spelling out "Step It Up" with flashlights. (PHOTOGRAPH BY MARY KATHERINE MCELROY)

empty cans and bottles to construct dozens of installations the night before commencement. The next day, parents and visitors heard a slightly boring speech—which gave them more time to look at the artwork (and message) around them.

THE PERFECT BACKDROP

Although holding a rally in a public park or on a street corner can be great, it's not exactly creative to hold all of your events in the same place. A group of folks doing something surprising in an unusual, visually arresting place makes people do a double take.

Take the example of the Water Rising Flotilla in Brooklyn, New York. Members of the Urban Divers Estuary Conservancy helped organize a demonstration on the banks of the historic and notorious Gowanus Canal, one of the most polluted waterways in the region, if not the country. The demonstration was followed by a so-called eco-cruise, during which more than a thousand community members and visitors bore witness to the human impact on the canal. The large group of paddlers ferrying down the fetid waterway grabbed the attention of passersby and the media—and their message was underscored the very next day when a huge rainstorm lifted the canal over its walls and flooded the basements of nearby homes.

Sometimes a place is important because of the power of storytelling. You know the special places in your town, the ones where couples go for walks and kids go to visit when they come home on break from college. The kinds of places that parents want their kids to be able to enjoy just as they did. The spots that feature in the private memoirs we each carry in our heads and hearts. They are the backdrops to choose for actions that you want people to remember for a long time.

Consider the "Take a Stand in the Sand" rally at Crane Beach in Ipswich, Massachusetts. "In my lifetime," wrote Franz Ingelfinger, the organizer of the event, "I have witnessed the virtual disappearance of the horseshoe crab from the shores of Crane Beach; my children stand to see those very shores disappear entirely." The backstory is that Franz's grandmother used to take him canoeing out in the estuaries and tidewaters surrounding Ipswich—it's *his* home and it has personal meaning to *him*. These areas, he points out, will all be open ocean, as will nearby beaches and towns, if urgent action isn't taken. "Together we can take a stand on the sand," he wrote in a powerful article in the *Boston Globe*. Don't hesitate to use a place that you care deeply about to get your message across—it's the most honest story you can tell because it's yours.

Creatively using a historical place as part of your event story will tap in to local sentiment. In Lexington, Massachusetts, residents, many in tricorne hats, held signs that proclaimed "Energy Independence Now!" and "Let the Energy Revolution Begin *Here*!" After the rally, a corps of fifes and drums led the crowd on a short march to the Minuteman statue, highlighting the importance of taking bold action against an enemy at least as nasty as the Redcoats.

PICK *YOUR* SITE

The first step in choosing a creative site for your action is brainstorming. Choose the most fitting public or private area and contact those in charge to attain the necessary permits as soon as you have it settled.

To avoid fines against your public-space-use permits, especially if you will be putting up a stage or a sound system or displaying or installing ad hoc art, you will need to be honest with officials. Your event should stay within the bounds of your permit.

For the most part, unless you're engaging in direct action, private areas will be off-limits for more creative events. That's not to say you shouldn't do a creative action if those in charge of the location don't agree with what you

PRACTICAL TIP

want to do, but it *does* mean recognizing what message your choice of location might send. For example, chaining yourself to a coal power plant will appeal to different people than rallying across the street from the plant on the town green—and you need to accept the risks and the trade-offs if you go forward with a nonpermitted action.

Once you've secured a location, try to meet with nearby business owners in person to let them know what you're planning. It will give them a chance to point out potential problems before they happen, and you'll probably end up gaining their support and involvement, which will help with spreading the word.

USE YOUR SENSE OF HUMOR

The scientists tell us there is nothing funny about climate change. In fact, climate change is so downright scary that some people just tune out when they hear about it. That's why "funny" can be so useful. You can use humor to your advantage as an organizer in ways that range from polar bear plunges with your neighbors to "culture jamming"—exercises like "redesigning" billboards—and quite a bit in between.

In the largest urban wilderness in the United States, Los Angeles's Griffith Park, a group of committed activists—and four clowns on stilts whose costumes included plastic cars and gas pumps—celebrated the Step It Up day of action by climbing more than two miles to 1,500-foot Dante's Peak, from where they looked out over the Los Angeles skyline. "Never underestimate the power of four stiltwalkers

Judith Lewis and Francis DellaVecchia (bottom row) with stilt performers Jessica Steen, Noah Veil, Mondo, and John Pedone (left to right), who climbed to the summit of Dante's Peak in Los Angeles, California. (PHOTOGRAPH BY DAVID NEWSOM)

where you don't expect to see stiltwalkers. Especially stiltwalkers wearing cars," wrote one blogger of the event. It was surprising, funny, and made for a great photo and media stunt. And, of course, they tied everything together with the message, "Why are we clowning around on climate change?"—a slogan that took on its most serious meaning three weeks later, when a drought-induced wildfire scorched the park.

"Radical cheerleading" is another way to add humor to your climate activism. In December 2005, an important meeting convened in Montreal on the future of the Kyoto Protocol amendment to the UN treaty on climate change. The U.S. delegation spent much of the week stalling other nations' efforts to discuss important next steps. Concerned students from around the world gathered at the Palais des Congrés, where all the meetings were taking place, to express their frustration with the inaction.

Early in the negotiations, several students who had registered as official observers for the conference began appearing as a cheerleading troupe. They dressed in bathing suits, Hawaiian shirts, and tank tops, brandished towels, and chanted what soon became a climate movement mantra: "Ooooh, it's hot in here. There's too much carbon in the atmosphere." They surprised delegates in the convention hall, bringing drama and fun to what would have been a dry week. On the last day of the meetings, the Montreal Marionette Collective set a festive mood with a massive papiermâché globe set in flames and a giant puppet of President George W. Bush playing a violin. While one group of activists stood outside the Palais proclaiming that the U.S. government was "fiddling while the world burns," others, dressed as members of the U.S. Congress, played real violins. Fun doesn't have to mean "lite." You can take nearly any message, image, or situation, whether it's a statement from a politician, a Peabody Coal advertisement, or a business conference, and twist it in a way that seriously embarrasses those you're targeting. You'll garner media attention and snickers of support from the general public.

Here's another great example, from the 2006 Los Angeles Auto Show. Two impeccably dressed men swaggered into the city's convention center, which was hosting the show, themed

"New Beginnings" and focused on new technologies, including high-mileage vehicles and "green" cars. Rick Waggoner, the CEO of General Motors—one of the least progressive auto companies in the world—was scheduled to give a speech highlighting the company's (belated) efforts to develop plug-in hybrids, hydrogen fuel cells, and more efficient vehicles. To ensure that his talk translated into real action on the corporate level, the two anonymous men—actually climate activists—took positions flanking the stage and, as Waggoner recited the last words of his address, stepped up to the podium to shake his hand and thank him. They looked like executives, so everyone assumed they were part of the show, including Waggoner. "We commended GM for offering an inspiring speech," said Mike Hudema, one of the activists, "and presented him with our pledge calling for higher mileage standards and carbon reductions. Sadly, he just tried to weasel his way out of our demand, stating, 'I think I'll let my speech speak for itself.' But we clearly said what the rest of the room was feeling—the crowd of journalists were laughing and applauding." The next day, a video of the incident showed up on the YouTube Web site and a number of blogs, and soon ABC, CNN, and Bloomberg News covered the story. The action was deemed a success!

In the end, the main reason you need to be creative is to keep the movement fun for yourself—and everyone else. We're talking about a grim subject, one that can seem so overwhelming that it leads to a kind of paralyzing despair. Music, art, good cheer—they give all of us the resolve we need to face it.

6

MAKE IT WIRED

Step It Up has been called the first large-scale "open-source" demonstration, a Web-based collaborative effort that allows anyone involved to add their own flavor to the project. We didn't want the day of action to exist solely on the Web, but we put an immense amount of time into figuring out how to use an array of digital tools that now make organizing far easier and faster than it used to be. Or, as in the case of Step It Up 2007, that make it possible—just two or three years earlier it would have been nearly impossible to coordinate and connect over fourteen hundred decentralized actions in ten weeks. The Internet has become the essential tool for building momentum behind the kinds of activism we describe, and there are some crucial things to understand as you put it to use in planning your campaign.

We would go further than that. Though the pace of technological change can be a bit overwhelming, we see the Internet as a way to help save our threatened planet. If a crisis of this magnitude had to strike at any time, it may as well be in the Internet age. Tackling global warming is going to require an unprecedented level of collaboration and communication at every level of society—and that's precisely why it's so vital that we learn to take advantage of the connections that the Net provides.

Organizing Step It Up was a crash course in how the Internet can be harnessed for social change. For three months, our small team was crammed into one room at our headquarters in Burlington, Vermont. Day after day, from this little corner of the country, we peered into our laptop screens and witnessed the self-assembly of a nationwide movement. As new groups came together to organize Step It Up actions in their communities, digital pins populated a digital map. Tapping in to the Web's nearly limitless communication potential, tens of thousands of people went online to post announcements, share ideas, debate tactics, exchange resources, and inspire one another with their passion and creativity. And in the days after the demonstrations, we used the Web to share a wealth of images and reports from across the continent.

The actions themselves didn't happen online—they were real-life, on-the-ground affairs, with neighbors coming together in the flesh to demand change. *We feel strongly that the Internet is best used to get people together face-to-face.* Too many organizations have put a blind faith in the Internet, thinking that simply having a basic online presence will immediately transform their group to a cutting-edge miracle of advocacy and activism. But to effectively harness the Web's potential, you must have a strategy to guide your work and a good set of tools to put your ideas into action.

When it comes to organizing online, a few key skills and some amazing new tools—we call them Activism 2.0—form the online organizer's toolbox and will help you put your principles into practice. There are five basic online elements you need to harness to support your group and your cause: the new "direct mail" of e-mail; the personal printing press of Web media; the networking power of the Internet; the borderless resources of information available on demand; and the borderless audience of people you can reach.

Disclaimer: With new tools constantly popping up, we're hesitant to recommend specific Web applications—we're writing this in June 2007, and there will be advanced versions of many of these programs even by the time this book is in print. So we've listed some solid tools that we think are going to be around for a while, though it's the ideas behind the tools that really matter.

BECOME AN E-MAIL GURU

Though it may not be the most exciting tool in the box, e-mail is nonetheless the cyberactivist's single most powerful weapon. You need to know how to craft compelling e-mails, send them out to many people, and handle large quantities of incoming mail. The ability to write compelling e-mails may be the single most useful talent an organizer can possess.

Keep Your Message Focused

When sending out an e-mail to a large group of people, keep it short and sweet—if you make it too wordy, people will simply click "delete." Try to keep each message focused on the very next action steps people can take.

Here's a sample e-mail we received from our friend Mike Tidwell, sent out to the Chesapeake Climate Action Network (CCAN), that embodies many of the best e-mail principles: it's punchy, specific, to the point, and eloquent. It draws the reader in and motivates him or her to action.

Dear Friend,

Ever wonder if group like CCAN, which tackles the global problem of climate change at a local level, is really necessary? Look no further than Sunday's Washington Post:

DC Area Sees Spike in Rate of Emissions
The Washington area is in the middle of a carbon dioxide binge, with emissions of this greenhouse gas from vehicles and electricity users having increased at more than twice the national rate between 2001 and 2005, according to a Washington Post estimate.
Read more.

What can you do about it? A lot. CCAN has been working tirelessly to educate area residents about the dangers of our carbon-heavy diet but we can't do it alone. Here's how you can help right now:

• **Everyone**—Support CCAN's critical work, become a member today!

Personalize it—either choose a salutation that will indicate your fellowship or invest in an e-mail service that allows you to assign a specific salutation to each recipient.

Link it—the simplest HTML to master is the link. Type <a href= followed by the URL, in quotes, that you want to link to, then >; after >, type the text you want the reader to see, and end the string of text with </a>.

- **DC Residents**—Speak up! <u>See below</u> for details on DC's first-ever Global Warming Lobby Day.

Ask for specific local involvement.

- **MD Residents**—<u>Write to Governor O'Malley</u> in support of a global warming solution for Maryland: reducing carbon emissions to 1990 levels by 2020.
- **Virginia Residents**—Va. leads the region in greenhouse gas emissions, emitting carbon dioxide at a rate nearly twice the national average. <u>Join us in urging Gov. Kaine to get serious about tackling global warming</u>.

With these new findings and the <u>crazy weather we've been experiencing</u>, we felt compelled to take action. Thanks for taking action with us and don't forget to check out April's highlights below.

Make it immediate— note the news.

Sincerely,

Mike Tidwell

Director, <u>Chesapeake Climate Action Network</u>

Mark your Calendar

<u>First Annual DC Global Warming Lobby Day</u>

Wednesday, May 23 @ 3:30

Wilson Building

Put the details after the snappy punch.

On May 23, join fellow DC residents in the first-ever Global Warming Lobby Day in DC history!

The DC Clean Cars bill has been introduced and now our council members need to hear from us. Let them know that climate change is a priority issue for all DC residents and that the Clean Cars bill is a step in the right direction.

Never lobbied before? Join us Wednesday, May 16th, for an <u>issue briefing and a lobby training</u>.

Break Up Your Text

Sometimes you'll need to convey a lot of information in a single message. In these cases, you should avoid long, stream-of-consciousness blocks of text. Instead, break down the information using bullet points, underlines, paragraphs, and bold formatting. Don't be lured into overdesigning your e-mail, however—a simple, straightforward note is often better than something with pixel-heavy pictures to download (and which can get your e-mail caught in spam filters) or slick-looking colored boxes (that may just show up as garble in some e-mail programs). But more to the point, this isn't advertising you're creating—you're talking about something deeply important and you want people to take notice of your words.

When we sent out our first Step It Up e-mail, we realized we would need a bit of room to get our message across. We kept each paragraph to three sentences or fewer—and some were just one sentence long. That's not the sort of writing most of us were taught in school, but it is direct and skimmable.

Dear Friend,

Invite and inspire.

This is an invitation to help start a movement—to take one spring day and use it to reshape the future. Those of us who know that **climate change** is the greatest threat civilization now faces have science on our side; we have economists and policy specialists, courageous mayors and governors, engineers with cool new technology.

Make it snappy— boldface your target—and make it credible—with just enough facts.

But we don't have a movement—the largest rally yet held in the U.S. about global warming drew a thousand people. If we're going to make the kind of change we need in the short time left us, we need something that looks like the civil rights movement, and we need it now. Changing lightbulbs just isn't enough.

So pitch in. A few of us are trying to organize a nationwide day of hundreds and hundreds of rallies on April 14. We hope to have gatherings in every state, and in many of America's most iconic places: on the levees in New Orleans, on top of the melting glaciers on Mt. Rainier, even underwater at the endangered coral reefs off Key West.

Describe your action—what are you asking from the person?

We need rallies outside churches, along the tide lines in our coastal cities, in cornfields and forests, and on statehouse steps.

Every group will be saying the same thing: Step it up, Congress! Enact immediate cuts in carbon emissions, and pledge an 80 percent reduction by 2050. No half measures, no easy

Describe your mission—what is your goal?

compromises—the time has come to take the
real actions that can stabilize our climate.

As people gather, we'll link pictures of the
protests together electronically via the Web—   *Reach for*
before the weekend is out, we'll have the largest   *collaborations.*
protest the country has ever seen, not in
numbers but in extent. From every corner of the
nation we'll start to shake things up.

By its very nature, this action needs all kinds of
people to help out. We can't make it happen—it
has to assemble itself.

Click here to sign up to host an action. We'll
coordinate the responses, introducing you to   *Ask for specific*
others from your area, and give you everything you   *involvement.*
need to be a leader, from banners to press releases.

**You don't have to have done anything like
this**—you're not organizing a March on
Washington, just a gathering of scores or
hundreds in your town or neighborhood.

We need creativity, good humor, commitment. If
you are active in a campus group or a church or a   *Evoke the*
local environmental group or a garden society or a   *opportunity for*
bike club—or if you just saw Al Gore's movie and   *fun.*
want to do something—then we need you now.
And by now, we mean now.

The best science tells us we have ten years to
fundamentally transform our economy and lead
the world in the same direction or else, in the
words of NASA's Jim Hansen, we will face a

"totally different planet," one infinitely sadder
and less flourishing.

The recent elections have given us an opening,   *Make it immediate.*
and polling shows most Americans know there's
a problem. But the forces of inertia and business-
as-usual are still in control, and only our voices,
united and loud, joyful and determined, can
change that reality.

Please join us. And please forward this e-mail   *Spread the word.*
far and wide.

Note that this e-mail does not shy from trying to
inspire—"take one spring day and use it to reshape the future." But it's also very careful in saying exactly what's being
asked of people and how they can go about participating. We
made sure to ask people to share the e-mail; the best way to
build your group, and the movement, is to do it friend by
friend. Striking a balance between nonintimidating humility
and confidence-building optimism also makes clear that you
want people to be involved and enjoy themselves.

CHECK AND DOUBLE-CHECK

Nothing makes you feel ditzier than sending out an e-mail to
three hundred people only to realize that you forgot to put
in the location of your meeting or that you asked people to
"Fight Global Worming." Avoid having to send out that embar-
rassing follow-up e-mail: make extra certain that you include
all the necessary information, and spell-check your work.

PRACTICAL TIP

Nail the Subject

An e-mail's subject line can determine whether or not people even open the message. You get fewer than a dozen words and you need to make the most of them.

Your subject lines should convey a sense of urgency, of fun, and of momentum. Imagine two different e-mails: one has the subject line "weekly meeting" and the other has the subject line "Help Create Anytown's Next BIG Action." Which e-mail immediately involves the recipient? Which one makes it clear that it's from someone—a person—in the community? Which one sounds like work and which one sounds like an important, creative project? And a word for the cyber-wise: there is a certain set of words that should be used only when absolutely necessary. "URGENT," "ALERT," "NEED," "DONATE," and their kin get old fast—especially if they are pasted on every e-mail you send, including that notice of your weekly meeting.

What never gets old is a personal touch. You're e-mailing this around *your* town. If you're well known, mark the subject line with a message title followed by "personal note from (your name)." If you're not well known, maybe someone in your group is. The best (or most available) kind of celebrity may not be a politician; look for someone—a minister, school principal, or local writer—who is widely respected in your community.

Bear in mind that too many e-mails will freak people out or, worse, fatigue them and get quickly sent to Trash without a look. Limit general mailings to one every couple of weeks, if possible. When the date of your action is imminent, it's okay to send e-mails more frequently, but focus those on new and essential information. You can always include links to a Web site with all the things people need to know if they have missed something, or include a summary at the bottom of the e-mail, after your sign-off. Watch out for adopting a crisis tone in all of your e-mails. "The Boy Who E-Mailed Wolf" is a cautionary tale for this age, too.

You should take every opportunity to build your e-mail list. Pass out a sign-up sheet at every possible meeting or event you pull together—and take that list wherever you go. If you have a Web site, give people a prominent opportunity on the front page to sign up for your e-mail announcements. It's absolutely vital to build the biggest list you can. Only a portion of the people on your list will show up at events, but the number will increase as your e-mail list grows. You're also educating your larger community via your announcements. And always—always—encourage recipients to forward the e-mail to as many people as they can.

E-mail Tools

WEB-BASED E-MAIL. When you're organizing anything, you're likely to get a boatload of e-mail in your box, and you're going to need a good way to manage it all. Gmail (www.gmail.com) is a free, Web-based e-mail service offered by Google. With nearly unlimited e-mail storage (2.5+ giga-bytes and growing), a clean interface, and powerful but easy ways to search, organize, and send e-mail, gmail is the best e-mail service out there for activists, especially if you need to be on the road and may not have access to specific e-mail software on other computers.

BCC (BLIND CARBON COPY). If you're sending out an e-mail to a large group of people, it's important not to put all of the e-mail addresses in the "To" or "CC" field. If you do, it leaves the e-mail list open to abuse for unrelated purposes and is sure to upset would-be supporters. Instead, use your own e-mail address in the "To" field and hide the addresses of the other recipients by putting them in the "BCC" field.

LIST MANAGERS. If you're trying to communicate with large numbers of people, using regular e-mail might not cut it. When your list gets bigger than a few dozen people, your messages can get flagged as spam by increasingly strict junk e-mail filters; your e-mail address might even be blacklisted. You might also want an easier way to control who receives your e-mails and how people on your list can communicate with one another.

This is when it's time to take advantage of a more advanced tool: the listserv. Listservs come in a few different flavors and at varying costs, but unless you are e-mailing thousands of people regularly, a free, Web-based service will do. You can set up an online group in no time at all. People will be able to subscribe or unsubscribe by sending an e-mail to a particular address, and you can easily control who gets to send an e-mail to the entire list. Add to that an ability to easily share files. And since these services work hard to make sure your e-mails get through, you're unlikely to be flagged as a Viagra-peddling cyber-delinquent. We recommend Google's list manager, Google Groups (groups.google.com), but you're certainly not limited to Google's tool. Yahoo! (groups.yahoo.com), Topica (lists.topica.com), and RiseUp.net offer similar free services.

If you need even higher rates of delivery and increased control (like seeing how many people on your list opened your e-mail in order to better estimate how many people will show up at a meeting), you might want to turn to a fee-based listserv. Though most people won't need these tools, two fee-based listserv services are relatively affordable and have distinct features that are particularly appealing to activist groups.

- ElectricEmbers.net: an activist-oriented Internet hosting service with fees set on a sliding scale. Several well-known environmental organizations use the system.
- DemocracyInAction.org: a not-for-profit that provides Internet tools for progressive organizations. Fees are determined based on the number of supporters and services provided. Among the handy features: e-mail forms that you can use to spread the word from the comfort of your own site.

MAKE *NEW* MEDIA

The next chapter is devoted to the art of seducing the conventional media—TV, radio, newspapers, magazines. But you know what? We had already organized 700 of the 1,400 Step It Up actions before the first major article about us appeared in any of those old school mediums. To do so, we made our own media and relied on the thousands of people who have leaped into the new media world in recent years. There's a vast array of user-created content, from blogs to podcasts to streaming video, that is increasingly important. Eighty percent of Americans are at least occasionally online.

In the last few years, the barriers that prevented people from generating online media have shrunk or disappeared completely. Now, with little time and less money, you can get up and running within minutes as a blogger, digital photo-journalist, or YouTube videographer. It sounds hip and trendy, but there's a catch: new media can't be used *simply* because they are "the new cool thing." Thinking strategically about if, when, and how to deploy digital media is just as important as having a grip on the technology itself.

Blog It

Blogs—short for "Web log"—are the foundation of the new, user-generated Internet. In their simplest form, blogs are easy-to-update Web sites where entries are displayed in reverse-chronological order, so that bloggers can provide news and commentary about issues and events as they unfold. There were about 80 million blogs worldwide in 2007, with that number doubling about every six months. Collectively, bloggers post more than eighteen updates per second, or about sixty-five thousand new blog posts every hour. In short, the "blogosphere" is an incredibly busy place, a buzzing universe of online interactivity.

Given these rather humbling statistics, you are probably asking yourself two perfectly reasonable questions: Why add to the online noise? And how will anyone ever find my blog? The short answer is that blogs offer you or your group an easy way to keep things lively, fresh, and engaging. They offer a forum for people to exchange ideas, post updates, solicit input, and request help—all without cluttering your e-mail box! Perhaps most important, they give human voices to your online presence. Only a relatively few blogs get many readers—the blogs that have something new, exciting, and important to say. That means you'll want to include some information in your blog that can only be found there, even for the people who studiously read every e-mail you send.

We made the blog at the Step It Up Web site one of the centerpieces of our campaign. We updated it several times a day, so that anyone checking in saw something new. And we learned a lot about what makes a good blog post. It should be fairly short, it should be about a discrete subject (plans for a particular rally, a word about a particular politician), and it should be written with confidence and just a bit of brio. It should take advantage of your collaborations—inviting people to write guest posts or post comments livens up the blog, provides an accessible space for fresh ideas and encouragement, and offers one more opportunity to involve everyone. Most of all, it should have a purpose—it should be designed to get people to think about new possibilities for a rally or campaign, or consider a new reason for joining in, or, often, just get readers to feel the momentum building. Here's an example of a blog post from our site.

Stepping Way Up to Create Aerial Art

Picture creating a collage out of little squares of colored construction paper (perhaps you remember something like this from elementary school). Now take that collage and expand it to an entire outdoor landscape. You hover eighty to a hundred feet above the ground in a crane, fire-truck ladder, or hot-air balloon. The little squares of construction paper become individuals either holding or wearing their particular color or pattern, whom you coordinate to make your collage. Now, connect this whole exercise to global warming in an incredibly powerful way involving education and community. And now, make it beautiful.

Daniel Dancer does this for a living. Art for the Sky [link] is a project designed as "a transformational adventure for schools and communities that forever changes our relationship to the sky." Recently, he has begun tackling the climate change issue in his projects with schools and communities, and he now has a project in the works in conjunction with Step It Up!

David Moen, a member of the team collaborating with Daniel, is currently doing work with the Oregon Zoo on Northwest Condors. He described the artistic connection between the condor image that will be created in mid-April and the threat of global warming:

"The Thunderbird, or spirit of the condor in most tribal cultures around here, represents the human relationship with the elements of change across the landscape—the elements that humans are not in control of, especially in weather, like storms and floods, those large impacting forces that bring disaster if not respected appropriately. Ring a bell?"

You bet it does, David. The art will represent the threat, but will also be demonstrating the solutions. Besides educating children and connecting to Step It Up, the activity is also "about seeing the whole for the parts from way up high and acting in relation to community identity instead of the individual." What more could one ask for?

We're looking forward to the event and beautiful art, guys. Thanks for Stepping It Up!

The best thing about blogs is that they operate at the speed of the Internet and are as much about links in as about links out. When we were putting together Step It Up, we were lucky enough to land an article about the campaign in the *New York Times*. The piece gave the site a small boost in Web traffic, but not the real surge in interest we expected (or hoped for). A day passed, and we figured our moment in the sun had come and gone. Then we got an e-mail from a supporter in New York City, who wrote, "Everyone I know decides what to read online based on which articles make the 'most e-mailed' list. Let's try to get the Step It Up article in the Top 10." A few minutes later, we put up a blog post pointing people to the article and asked them to e-mail it, via the *Times*'s "e-mail this" tool, to as many people as they knew. People responded: within a couple hours, it had shot to the seventh most e-mailed article and our Web traffic soared. Many local online news sites provide the

In conjunction with the release of *Everything's Cool*, a film about global warming that premiered at the Sundance Film Festival in Park City, Utah, John Quigley of Spectral Q organized a high-impact aerial art piece, spelling out "Step It Up" and "We hear your call" in Inuit, a gesture to a similar aerial art project created in Iqaluit, Canada, in 2005. (PHOTOGRAPH BY CHRIS PILARO)

same feature, and most blog-hosting services allow you to build in "track-backs"—a listing of other Web sites that have linked to your blog post. Track-backs go both ways, so your site will likely get linked quickly. And the more meaningful the links you make, the more traffic you'll get—links are one part of how your Web site ranks in many popular Internet search engines.

Play Tag

Another way people will find your blog is through "tags," self-made indexing words that you can attach to your blog post, or other Web tools. You might think of tags as categories—some of your blog posts might be about "action ideas," others about "community leaders," still others about "Congress" or "data you can use." Putting tags on your blog posts allows people to find all the posts on "data you can use" with a single click.

Tags become really powerful when they spread beyond your own Web site. Sites like Delicious (http://del.icio.us) allow people to publicly bookmark a list of Web pages they want to find again, attaching tags so that they can track down the pages in the future. The tags don't have to match the ones you assign on your blog—they can be anything. So if you set up a "permalink" (or standalone, linkable) page for each post through your blogging software, you can also "remember" the page on a Delicious account and add "big picture" tags: global warming, climate change, your town, your event name. It helps to think about what other people might be using as tags for similar Web pages. If someone is bookmarking an article about the Intergovermental Panel on Climate Change report and you have a blog post about it, what tag might you happen to have in common? When people bookmark a Web page, they can see other Delicious users who have also bookmarked it.

When people attach a tag to a bookmarked page, they can find other bookmarked pages with the same or similar tags. Delicious users can also search for tags.

Delicious is just one of many user-generated bookmarking sites, including Fark, NewsVine!, and bookmarking services at Yahoo! and Google. Also, Digg.com and Reddit.com allow users to rate pages posted by others, so those that are the most interesting to registered subscribers rise to the top, whether they are blog posts or articles from the top news sites in the world. More such sites helping people navigate the seemingly infinitely expanding blogosphere crop up each month.

THE VALUE OF THE VISUAL

In an incredibly short amount of time, online photo- and video-sharing sites like Flickr.com and YouTube.com have turned the entertainment industry on its head. All of a sudden, everyone is a creator—and fewer and fewer people are pay-for-play consumers.

But do video and photo sharing have value for activists? Sometimes. Visual media pack a potent emotional punch that can't be replicated with text alone—not even with a thousand words. Documenting your efforts with photos is, more than anything else, fun. It's a great way to keep people involved, and your photos will tell an ongoing visual story for your projects and your group. Digital photos and videos are also amazingly easy to distribute. For Step It Up, we planned from the start to use photos to display the breadth of the day's activities. By the time night fell on April 14, we set up a collection of pictures from all over the country in a streaming slideshow on the Web site; the show united all the hundreds of groups with one common visual story. Someone who had no idea what Step It Up was about could look at the pictures and understand within seconds the passion, creativity, goals, and message of this nationwide call to action, especially since we asked everyone to include an "80% by 2050" banner in their photos.

Similarly, online video can be creatively used to personalize and share your message, drum up excitement, publicize events, and get attention. In New York, organizer Ben Jervey had ambitions to stage a massive Step It Up rally and interactive art project called "Sea of People." Ben's idea was that thousands of people dressed in blue would line up in lower Manhattan to show where the new tide line is forecast to fall if global warming continues at its current pace. A week before the event he realized that a crucial piece was missing: all the *people* he needed to make the *sea*. He didn't have an army of volunteers to put up flyers or the funds to place high-profile advertisements, so he turned to online video instead. He got his friend Blake to whip together a quick, inspirational promotional clip for Sea of People and uploaded it to YouTube. With a portable digital video recorder, Blake taped three people in three nearby locations in Manhattan as they described sea level change and the event. A few cuts between each person in a video editing program (for example, iMovie) plus a simple background soundtrack and a frame shot of the event details at the end and they had a world-class outreach tool. Within hours, people were e-mailing the Sea of People video to friends and posting it on blogs. You can even make movies

The Lower East Side Girls Club of New York expressed its concerns for future generations at the Sea of People rally in Manhattan. (PHOTOGRAPH BY LYN PENTECOST)

from a series of digital photos or by using the movie setting on many digital cameras.

Video can also be a great way to share information with folks within your organization. When April 14 was just a few weeks away, we started worrying that people weren't exactly sure of the basic elements for organizing an action. We sent e-mails and created an "organizing guide," but it seemed some folks were still missing the details. In our fantasy world, we would have magically teleported around the country, checking in with every local organizer face-to-face. Since that option wasn't available, we borrowed a camcorder and made an instructional three-minute video that talked through the organizing process. The whole thing was shot, edited, and posted on YouTube within a few hours. Within two weeks, thousands of people had watched it online. We received countless e-mails from people who said that they were on the fence about pulling together an action until they watched the video—which in three snappy minutes made them realize they were up to the task and could have fun to boot.

New Media Tools

BLOGS. Like everything else on the Internet, you have a lot of options when it comes to blogging. The tool that is easiest to set up (but also the least customizable) is Blogger (www.blogger.com), which can get you up and running with your own blog in a matter of minutes. Blogger offers a range of design templates (with some customization options) and the ability to add labels (or tags) and moderate comments. You can also give readers permission to "e-mail post" to friends, something you should take advantage of. If you need more functionality and flexibility, check out WordPress (www.wordpress.com) and TypePad (www.sixapart.com/typepad), which have feature-packed subscription versions. Using a plug-and-play blogging tool can also save you the expense of buying an Internet domain name, which may be overkill for some short-term, snappy campaigns.

PHOTO SHARING. Web sites like Flickr (www.flickr.com) allow you to create a centralized online repository of a large number of digital photos and to contribute photos to public "pools," or groups of photos around an event, a theme, a city, a neighborhood, or anything else you want. Flickr users can organize and tag photos, too. You can easily embed photos on your site or other Web sites and blogs. There are even Flickr tools to embed a slideshow of photographs (www.flickrshow .com and flickrslidr.com). Flickr has adopted a collaborative approach to online tools and encourages users to develop them, so keep an eye out. Other free photo-sharing sites include Photobucket (www.photobucket.com) and Twango (www.twango .com), which support digital audio and video files as well.

Some Flickr users upload photos, under a use license developed by the not-for-profit Creative Commons that allows you to use them free on your Web site or blog (sometimes with a credit to the photographer and always with a credit to Flickr). To add visual appeal to a blog post, simply visit the advanced search function on Flickr and look for photos with a Creative Commons license; if you want a picture of your local town or of the Arctic, you will of course have to search for the appropriate tag, too. If you don't find what you're looking for, you can always turn to a stock photography source, such as iStockPhoto.com, which carries a database of tagged photographs that are available for use on your blog for a fee.

VIDEO SHARING. There are only a few big players in the game of online video, with the most dominant being YouTube (www.YouTube.com). After registering on their site, you can easily upload videos, create a YouTube channel, and embed any video in your Web page or blog. YouTube also allows you to tag videos. If you want to check out other options, we suggest Vimeo (www.vimeo.com) and Google Video (http://video.google.com); both have a higher file-size limit for video uploads than YouTube.

DIGITAL VIDEO DOS AND DON'TS

If you decide to create digital videos to get word out about your event, follow a few handy guidelines for making it effective as possible.

- Avoid the talking head. There's nothing less interesting than watching three minutes—or even thirty seconds—of a person talking at their desktop computer's webcam.
- Avoid the roving camera. Take steady, easy-to-watch shots that can be spliced with other video using quick cuts in your editing software.
- Choose lively locales. Let the background in your video provide some visual interest and play up your local angle.
- Invest in audio. You don't want poor sound quality to distract from your message. A good-quality stand-alone microphone can be bought for about $15.
- "Brand" it. By that we mean linger on an easy-to-read shot of the event details or message and tag it with your organization name.
- Don't have a camera? Record a lively audio—particularly if you can include music—and create an audio file or "podcast" that can be embedded on your Web site and posted on music-sharing sites.
- Focus on snappy fun. That's what makes a video go viral.

AVOID "ONE-CLICK ACTIVISM"

Gotten an e-mail from a big organization lately? Did they ask you to "take action" by signing an online petition or contributing money? We call that one-click activism, and not only is it a limited use of the Web, it squelches as much momentum as it creates.

The good news is that there is a better way for concerned citizens to harness the Web: decentralized networking. And decentralized networking is exactly what we need to fight global warming, that is, a sustained and lively social movement, not a movement that can be bought with money. The movement to stop climate change is being spurred by open dialogue and collaboration between individuals, the kinds of interactions that give rise to communities that are willing to do more than spend thirty seconds signing an e-petition. And perhaps above all else, it's a movement that has room for everybody to be an active participant and contributor.

That means we've got to actively create vibrant networks—spaces on- and offline for individuals to connect in meaningful ways. Rather than restricting communication to messages flowing up and down (as in a traditional hierarchy), we have to allow people at the "ends" of our networks to connect and information to flow in every direction. For the Step It Up day of action, we made sure that individuals across the nation could connect to each other online in our "organizer forum," a chat room and discussion board on the site where people could brainstorm and tackle problems together. In our experience, many minds are far better than a few.

We also discovered that it's easier to get people out on the street with sign-up forms, educational sheets, event flyers, posters, stickers, and other old-fashioned activist stuff if they can download it to their computer on the fly and at their convenience rather than having to schedule an appointment to pick it up from a central location. Try to stick to formats that most people can print at home—that means offering black-and-white as well as color versions, and plotting out stickers following the layout of a specific type of easily purchased mailing labels (and letting people know which one you're using). Adobe Portable Document Format (PDF) is especially helpful if you want to include logos and other artwork and still keep your file sizes relatively small. The software needed to make, view, and print PDFs is available free on the Web.

To keep things collaborative, solicit and accept ideas and designs from as many people as you can and encourage people to share their printable materials on your Web site. To involve as many members of your community as possible, invite people to translate educational sheets and flyers. You can even run a poster-design contest online and let everyone vote for (and download, if they like) their favorites.

Digital Networking Tools

INFORMATION AND RESOURCE SHARING. GoogleGroups, discussed above in the "E-Mail Tools" section, is a fantastic resource for opening up channels of communication and sharing files online. A Google-hosted public calendar, Leapfrog Into Action, got word out about Step It Up and other environmental and activist events. A tool called Backpack It (www.backpackit.com) allows users to share to-do lists, files, notes, calendars, and more, which can be very helpful when you're putting together a short-term project with a lot of people.

SOCIAL NETWORKING. The rise of social networking tools, designed to help individuals create active online communities, has been meteoric. While no social network is tailor-made for local action groups, they are evolving rapidly and are yet another way to let your friends, and their friends, know what you're up to. Though MySpace (www.myspace .com) and FaceBook (www.facebook.com) are the prototypical social networking sites, scores of others are out there that are worth investigating. Check out Orkut.com, Hi5.com, Care2.com, and Zaadz.com.

ONLINE TO OFFLINE CONNECTIONS. We've said it before, and we'll say it again: there's nothing like getting together face-to-face. Web sites like Evite (www.evite.com) and MeetUp (www.meetup.com) allow people to plug their e-mail lists into event management software that helps you announce, schedule, and publicize offline gatherings. You can also search MeetUp for gatherings in your area that draw people who are interested in environmental and other issues.

STAY INFORMED—AND INVOLVED

The great promise of the Internet was free and open access to information. Well, now we've got it—for anyone with a connection to the Web, an entire digital universe is only a few keystrokes and mouse clicks away. We'd be foolish to forgo the opportunity to stay informed and involved online—indeed, keeping abreast of the latest developments is vital to any campaign's success. But there's so much out there, it's tough to know where to begin and it's easy to get overwhelmed. Not to worry. With the right tools and resources, you can have the information you want, from all around the world, delivered to you in no time at all.

Information Tools

ENVIRONMENTAL WEB SITES. If there's one hub that environmentalists rely on for the latest news, it's Grist (www.grist .org), the Seattle-based daily online news site focusing on all things green. Funny but also impeccably accurate, it's a key place to get your message across (see especially the Gristmill blog) and to get a sense of what others are up to.

Other great sites include Treehugger (www.treehugger .com) and World Changing (www.worldchanging.com), not to mention It's Getting Hot in Here (www.itsgettinghotinhere .org), the hub of the youth climate-organizing universe. For references on everything scientific, don't miss Real Climate (www.realclimate.org), a site written by working climate scientists (no politics or economics involved). The Resources section in the back of this book includes a list of some key environmental Web sites.

NEWS ALERTS. Have a particular topic that you need to keep track of? Create a Google news alert (www.google.com/alerts) for a given term—for instance, "global warming"—and you will receive scheduled e-mails (your choice of breaking, daily, weekly, monthly) that list how and where that phrase is used in online newspapers, blogs, and more. News alerts are a great source for blog posts, too—you can highlight small, under-the-radar articles, note connections between stories or to your local community, or provide a handy digest for people on the go. Which brings us to . . .

RSS FEEDS. Too many blogs, headlines, and Web sites to keep track of? Many sites now offer RSS feeds. RSS stands for "Really Simple Syndication," which essentially means that you can look at condensed versions of Web sites or blogs all in one place, sort of like a neatly packaged table of contents. You can view RSS Feeds with a "feed reader" or "feed aggregator" such as Google Reader (www.google.com/reader) or BlogLines (www.bloglines.com). RSS feeds are a huge help if you're like most activists—frantically trying to keep up with a million things without a moment to spare. And, of course, you can also give the readers of your blog an option to add your site to their RSS feeds.

WIKIPEDIA. A completely free, open-source online encyclopedia, Wikipedia (http://en.wikipedia.org) can be edited and added to by anyone. The result, amazingly, is a thorough, constantly updated treasure trove of link-packed information that has proven to be remarkably fact-checked—and search-engine friendly. Whether you want to research your senator or carbon sequestration, Wikipedia, with nearly two million individual articles, is usually a pretty good place to start. And if you are an expert on a topic not covered already, you can add it!

CRAFT YOUR IDENTITY

A Web site (or a blog or a YouTube video) isn't worth creating if nobody is going to see it. Beyond the ideas mentioned so far, entire books have been written about how to attract people to your Web site, and the fact of the matter is that there is no tried-and-true method for having your Web site "go viral"—meaning hit the jackpot in terms of Web traffic. But all activists should ask a few questions as they develop an online presence.

IS IT USEFUL? This is the paramount rule of online organizing: if there's no reason for people to check out what you have created online, you won't get traffic. Give people relevant things to do, use, and read online and they'll keep coming back.

IS IT SIMPLE? An overly cluttered site that tries to do too much will turn people off, and they'll quickly surf away in confusion. You can't be the one-stop shop for global warming, so it's best not to try. Instead, keep things streamlined, readable, and highlight the most relevant information about your actions and your community.

IS IT FUN? Don't just post a text-based calendar of events—post pictures from the local foods potluck you just had. Don't just link to an article, embed a YouTube video that helps to tell the story.

IS IT FRESH? If there's never anything new on your Web site, people will stop checking it out. Keep things updated with local news, reports from ongoing projects, updated calendars, photographs from recent events, relevant thoughts, your late-night manifesto, links to interesting blog posts, etc.

IS IT DISTINCT? Without a distinct feel, voice, and purpose, your site is destined to languish in cyber-obscurity forever. Make a logo for your project or your group and post it on every page. If you are creating a Web site for a local global warming action group, post profiles of your members and write about local threatened areas and potential solutions to the problem. For Step It Up, we swallowed our natural shyness and posted lively, brief bios of ourselves ("my whole life in 250 words or less"), with photos, even—and organizers around the country said it was easier to connect with us when we weren't just names.

IS IT LINKED? The greatest single thing you can do to drive traffic to your Web site is to get links to it everywhere you can imagine—short of spamming the blogosphere. Add a link to your e-mail signature. Get others to post links to your site on their blogs, and do the same to generate track-back mentions. Include your Web site URL in all printed materials. And don't be afraid to get creative—a friend of ours spent a cold winter night barhopping while wearing a tank top and a giant sign that advertised the name of her organization's new global warming blog, It's Getting Hot in Here. Her advice: do whatever it takes to get your online presence noticed.

We want to reiterate one thing in closing: it's easy to get so caught up working on the Web, creating beautiful images, and linking to magnificent sources of information that you forget to organize. (Also, it's less scary to do things online, because no one ever turns you down—they just ignore you.) Being wired is not a substitute for actually making contact with the people in your community that you need to persuade; it's one more tool for making it easier. Use it wisely. From now on, your tools, your network, and your movement are just a few mouse clicks away. So as you're organizing online and in your community, be sure to send us an e-mail to let us know how it's going. You can always reach us at organizers@stepitup2007.org. From now on we're all connected.

7

MAKE IT SEDUCTIVE
(to the Media)

Even if you do everything absolutely right there's no guarantee that you'll get a lot of coverage in the press. That's because the news is inherently unpredictable—a big news story completely unrelated to global warming could break the same day as your action. If you don't get much coverage, don't fret—you'll have all the good and lasting effects of educating the people you reach directly, no matter the news.

Yet getting press coverage is worth the effort since it can multiply the effects of your hard work and gives everyone involved a nice boost. And you can vastly increase the odds of getting coverage if you understand how reporters and editors think—what it is that makes them want to write about or film a story, and then to give it good placement in print or on broadcasts. With Bill's lifetime of experience as a newspaper and magazine writer, Step It Up had an insider's insight into how the process works—and you will, too.

THEY DON'T CALL IT A NEWSPAPER FOR NOTHING

The first thing to understand is that reporters and editors are deeply interested in what's new. And by contrast they're deeply uninterested in anything they perceive as old—as yesterday's news.

When we started planning Step It Up, many people suggested that the climate movement needed a massive march on Washington. We thought it was the wrong strategic idea for many reasons, including the mixed message it would send to have people crossing the continent spewing carbon behind them in order to protest global warming. But we had also noticed that recent big demonstrations on the Mall in Washington or in New York's Central Park had received virtually no news coverage even though they had drawn hundreds of thousands of demonstrators. The frustrating reason: reporters and editors saw them as "old hat," as a cliché. Take the early 2007 antiwar protest that successfully mobilized tens of thousands of people in the capital. The *Washington Post*, one of the few newspapers to give it any coverage at all, emphasized what it saw as the "newsworthy" part of the event: a few hundred militants who skirmished with police. So we decided that a dispersed action all across the United States would seem "new" and generate more coverage, even if it drew no more total people than a big march on Washington. That proved right.

The same dynamic is probably at work in your city or town. If it's the usual suspects doing the usual thing in the usual place, there's not much chance the press will show unusual interest. You should think beyond the standard rally or march for something that is distinct enough to generate media interest. (This is a good idea for more than attracting the press—it's also a good way to attract new people to the global warming movement.) A starting place would be chapter 5, "Make It Creative," but also consider how to take advantage of your place, your history, and your struggle in ways that entice media coverage.

THE POWER OF PLACE. In Helena, Montana, local organizers Becca Leaphart and Ben Brouwer were planning a march through town, but they ran into problems getting permits—which turned out to be a blessing. When they were forced to think a little harder, they came up with an inspired symbol as the centerpiece of their action: the town's historic fire tower. For years it had housed a fire spotter, who protected the forested community from wildfire; the tower was known as the "Guardian of the Gulch." You couldn't ask for better symbolism for a global warming rally—especially when they recruited a city commissioner to help them get permission to ring the tower's bell, which many in the community had never before heard. Becca and Ben handed the first paragraph of the story to their local news reporters.

THE POWER OF HISTORY. Adi Nochur and his co-organizers in Boston wanted to highlight the danger of our dependence on coal. The city did not have much visible connection to the coal industry, but Boston is a place rich in revolutionary history. So as part of an international day of climate action in

More than a hundred residents of Helena, Montana, gathered at the town's iconic fire tower, nicknamed the "Guardian of the Gulch," to draw awareness to the threat of more frequent wildfires. (PHOTO-GRAPH BY KATIE KNIGHT)

2006, Adi and his friends staged a Boston Coal Party, dressing up in colonial garb and dumping coal in Boston Common in the spirit of America's tea-dumping ancestors. The innovative street theater powerfully connected the city's special history with the dirty-energy issue, and hundreds of people showed up to watch the festivities and participate.

THE POWER OF STRUGGLE. People have an inherent interest in stories of struggle and success. Walking fifty miles is challenge enough to raise some drama—a reason our 2006 Vermont march raised attention was that by day five the TV cameras could take pictures of our blistered feet as we rested by the road. For Step It Up, a hardy crew of skiers ascended one of the high peaks in New York's Adirondack Mountains in the midst of a blizzard. The picture of them fighting their way to the top with a Step It Up banner dominated the front page of the *Albany Times Union*, upstate New York's most important paper, and was picked up by scores of online news services.

HOOK THEM WITH A STORY LINE

Reporters love a narrative, a story line that lets them understand why something is new and different. If you can provide that narrative for them, if you can almost write the story yourself, you'll get covered. That requires figuring out the angles that might interest a reporter and going far beyond "we're having a rally and we hope a lot of people come" and "global warming is really, really dangerous." Here are a few tried-and-true narratives that you can adapt to your situation.

THE SUPERLATIVE. Find a way to boast about your action. Is it the "first interfaith gathering in the area on global warming"? Is it "the longest march in the Bay Area in a decade"? Is it "the biggest piece of aerial art ever on a North Coast beach"? Don't make outrageous claims ("perhaps the biggest" and "among the first" are useful phrases when you can't prove your superlative without a doubt), and if someone points out that there was a longer march the year before, thank them enthusiastically for letting you know—and think of something else to boast about.

DAVID VS. GOLIATH. There's a temptation to make an organization sound bigger and more polished than perhaps it is—we figure people will take us more seriously the more professional we seem. But that's not always the case. With media, it can be an asset to appear a little amateur, a little homemade. Reporters and editors are used to dealing with organized interest groups with slick press releases and often find it refreshing to get a different pitch. In discussing Step It Up, we usually emphasize that we were six college kids and a writer, and hence a little clueless. (See, as evidence, the first paragraph of the introduction to this book.) This had the advantage of being both true *and* interesting, and it made the success of Step It Up more remarkable. So invite reporters to see that you're meeting around a kitchen table; let them know that you have organized four hundred people on sixty-three dollars. Keep in mind, though, that this approach works as a news story only if David is beating Goliath.

STRANGE BEDFELLOWS. One of journalism's favorite narratives is the happy odd couple. And it never seems to grow old, no matter how odd. Mine the collaborations you've built to find odd-couple pairings that will appeal to the media, from the hunter who stands alongside a conservationist to members of opposing political parties. Are you an evangelical pastor concerned about global warming? Invite a secular environmentalist to share your pulpit some Sunday before your event, and let the local newspaper know. Are you a secular environmentalist? Find the local evangelical pastor who shares your concerns, and make him or her a featured speaker at your rally. In Galesburg, Illinois, organizer Isaac Yowder said he got a great deal of interest simply because he worked at a meat-packing plant. "Some people joined just because they couldn't believe I was doing such an unusual thing," he said. It's just the kind of story a reporter would be able to turn into a fine feature. In Tampa, Florida, employees of Sam's Club were key to the Step It Up event, setting up the tables and purchasing bottled water for the day's demonstration. The combination of big-box store and climate activism was odd enough to warrant media interest.

THINK DRAMATICALLY

In chapter 5, we shared many different methods for dramatizing your message. This is especially crucial when it comes to the media, because from a reporter's point of view there are a couple of problems with global warming. For one thing, it happens more slowly than traditional news events. Its effects are also felt everywhere, instead of in a particular place, that is, your hometown aka your media market, which is how media people are trained to think about the news. In a sense, climate change is too big a story, so your job is to break it down to reasonable size.

One way to do that is with the latest forecasts for what climate change will do to your area. Many good newspapers have run series about such possible impacts, and someone at the local college or university will probably know where the latest data on your region can be found. That's good—but then you need to bring it to life.

How can you make it dramatic? Organizers in Jacksonville, Florida, hired a crane and suspended a yacht twenty feet in the air to show how high the local sea level would rise if the Greenland ice shelf slid into the ocean. That hit home for everyone who saw it. (And they set the crane up in front of the football stadium for extra hometown appeal.) In the previous chapter we described the Sea of People who showed where the new tide line would fall in lower Manhattan if the ocean started to rise. Think about these two actions: anyone who caught a glimpse of them suddenly had a visceral sense of how global warming would change the neighborhood.

And people *would* be likely to catch a glimpse of them precisely because they are the sort of actions that appeal to an assignment editor dispatching a newspaper photographer or a TV camera crew. The point of a great visual image is not that it's pretty (though that helps) but that it carries meaning: if the correspondent is doing a stand-up underneath a boat hanging in the air, a lot less explaining has to be done to get the point across. And since he or she has only forty seconds, maybe, for a report, anything that requires too much explanation simply doesn't get covered.

Feel free to borrow ideas that others have used elsewhere— until something becomes clichéd, there's no prize for originality, so a crane hoisting a yacht will work in any coastal community until one of the events gets a national media hit. Think about the striking visuals that define your community. In Williamstown, Massachusetts, a village of perfect New

Step It Up activists snorkeled and scuba dived off Sand Key reef in Key West, Florida, part of North America's only remaining living coral barrier reef. (PHOTOGRAPH BY CRAIG QUIROLO)

England quaintness, the white church steeples along the main street are the thing you remember most about the town. How inspired, then, to take a big red banner and turn one of those steeples into a giant thermometer with the mercury rising. In Mobile, Alabama, on the Gulf of Mexico, organizers found a polar bear costume, then added a bikini. "That's her normal attire now," David Underhill, chair of the local Sierra Club chapter, told reporters. "She's bearing a message from the melting north." (Pun intended.) It takes a pretty hard-nosed city editor to pass up a picture of a polar bear in a bikini.

There are different kinds of drama, including some you may want to avoid. Angry people are inherently dramatic, because you never know if things will get out of control. For the moment, though, we think the climate movement has less to gain from people being angry than from people pointing out that it's completely reasonable and obvious to take steps against global warming. The day may come when anger is the appropriate emotion—but in relation to the press, understand that if there's confrontation in your event, that's the drama that will dominate the coverage. And it will drown out any other message, so make certain it's *the* message you want to get across.

Of course, sometimes stuff happens that is beyond your control. In Saratoga Springs, New York, a huge throng gathered downtown for Step It Up. A solitary antiwar protester left his regular vigil in front of the post office to see what was going on—and the photographer for the local paper put his picture on the front-page story about the rally, simply because his placard was the most aggressive. The organizers might have liked to see a different picture, one that emphasized the issue of global warming more squarely, but think of it this way: at least it got on the front page.

Which leads us to the most important piece of advice about the media: when everything is over and done with, it's largely outside your control. Don't worry too much. Good and bad images and stories are transitory, gone in a day or a week, and definitely by the time you start brainstorming your next action.

CREATE A RELATIONSHIP

Now that you have your story pitches down, it's time to turn to the care and feeding of your media list. Reporters and editors are extremely busy, but they are also much more community minded than you might expect—especially local reporters and editors (though the same principles apply at all levels of the business). Your job is to be helpful to reporters—to provide them with information without being pesky.

The temptation when dealing with reporters is to send a press release and leave it at that. But press releases drift into newsrooms like snowflakes in a blizzard. It helps to get to know a reporter and an editor early on in your work. Say you're planning a rally two months hence. Call or e-mail and ask if you can meet with the editor for ten minutes—chances are he or she will say yes. Journalists want to know the people in their communities who will be making news, and they like to have a sense (or a sneak preview) of what's going on.

GET YOUR TEN MINUTES

What seductive words will get a reporter or editor to accept your request to meet? You'll need to craft a two-sentence (or *maybe* three, if you include your ID) query that makes the most of your news and your narrative. You have about twenty seconds to ask for your ten-minute meeting. Try something like these.

PRACTICAL TIP

- "My name is John Smith and I'm a lifelong resident of Anytown. For April 14, I'm planning what's shaping up to be the biggest rally in our town in two decades and I would like to meet with you for ten minutes to tell you about it."

- "My name is Jane Smith and I own the sporting goods store on Main Street. I'm part of a community group that will be demonstrating how global warming will affect our town by 'kayaking' down the new coastline that will run right in front of my storefront. I'd like to give you a sneak preview of our plans over ten minutes on Thursday."

Once you're in the office, don't use your ten minutes to explain why global warming is a bad idea. The editor of the local paper or the news director at the radio station cares less about global warming than he or she does about the politics and news of your particular town. Instead, lay out the basic plan for your campaign—the things you plan to do in the lead-up to your event, the kinds of people you have involved so far. If a few things in your early planning seem unusual—and thus newsworthy—mention them; they might spark interest in a feature story. Consider how the people who would be featured would help pull in an audience for the newspaper or station. Perhaps you've got a group of residents at the local nursing home who have volunteered to handle the clerical chores or a group of schoolchildren who will be performing at the event—that will get their families reading, clipping, watching, and video recording. But this isn't the time to press for commitments on stories—all you're doing is establishing a relationship and demonstrating that you're accurate, helpful, and not so self-righteous that they'll need to roll their eyes when you call with an update.

A place you should not neglect is your local office of the Associated Press. The AP sends out stories along the news wire to local papers throughout your region. If you can get AP to cover your action, its story will be a real bonanza—widespread coverage in far-flung areas where you won't have the connections or time to be working the newspapers as you do close to home.

CHOOSE A SINGLE POINT OF CONTACT

You should identify one person in your planning team as the main contact for the press. That person will hold the initial meetings and build the relationships with reporters and editors, and he or she should have the ability to be informative without being boring and persistent without hectoring. As far as the media go, your official contact is the public face of the event. This doesn't mean the person needs to be an expert, but he or she should be self-assured, optimistic, confident, and convincing. Your media contact should also be easily available; a person who checks e-mail only weekly is no doubt virtuous, and probably saner than the rest of us, but should be given a different job. Having enough sense for how journalism works that the person can make the lives of editors and reporters easier also helps—at the very least, your contact should be prepared and willing to provide interested media with the phone numbers and e-mail addresses of local experts who can be interviewed (which requires, to some extent, nurturing good relationships with the experts as well).

Everyone else on your organizing team should back off from calling reporters and editors directly. The last thing you want is for six people each to call the same reporter, who will feel besieged and drop the story rather than sort out the right contact for your event. But that doesn't mean everyone else won't play a role in helping to ensure good coverage.

WATCHING THE BYLINE

Want another way to build a close relationship with a reporter or editor? Track the types of stories that your local newspapers and radio and TV stations like to cover that fit the narratives you'll be pitching for your action—and think beyond the science, environment, and political beats. Who at the local affiliate tends to be on camera when the station goes to the retirement center to do a community feature? Who likes to cover stories of ordinary folks trying to fight the establishment? Does your local paper always cover any event that happens at Central High? If you know who reports the sort of story you want to place or what sorts of locales get regular coverage, you'll do a better job of caring for and feeding your journalist.

MANUFACTURE MEDIA MOMENTUM

People like to be involved in things that promise to be successful, and reporters are people, too. One of the ingredients of a successful campaign is the impression of "snowballing"—that your supporters and participants are numerous and growing. Much of the work you do to build a crowd for your action will build interest among the press. And vice versa.

An early step is to land your first article in a print publication, and a good place to start is the alternative paper in your area—they're likely to be well disposed to what you're doing. Ditto the community radio station, if there is one nearby. Once you've got that clip or those tapes, send copies along to your contacts at the more mainstream press. You can do the same thing with blog posts—collect a few about your action and forward them to the reporters you're working with. In general, press begets more press. Reporters are more comfortable running with a story if someone else is covering it. You can bootstrap your way up the media food chain in this way.

But you don't have to rely solely on getting people to write stories about you. The local press is wide open to uncensored input from the outside. Bill's father was ombudsman for the *Boston Globe*, and Bill learned a lot from him about what people actually read in newspapers. Number one was the comics—if "Beetle Bailey" was inadvertently left out for a day, the phone rung off the hook. But letters to the editor were close behind. Especially in a small community, they're what people turn to in order to take the temperature of the town in a given week.

Writing a good letter to the editor isn't hard, and it needn't be long. It isn't necessary or even useful to recite the whole story of global warming; assume people know. Instead, connect to something local, and then get the information out about your event.

> Dear Editor:
>
> I was walking on the beach at Point Reyes last Sunday after reading the story in your paper about the possibility of sea level rise from melting ice shelves. I calculated that the twenty-five-foot rise in sea level would be enough to submerge Highway 1—that's the world we may be leaving our kids.
>
> Some of us here in Marin are trying to do something about it. We'll be holding a protest on Saturday, June 1, to urge our representatives in Washington to pass legislation that would cut carbon emissions 80 percent by 2050. A team of hang gliders towing banners will descend from Mt. Tamalpais, and windsurfers with the same message on their sails will be playing in the waves. We'll all converge on the beach at noon and form a human postcard to Congress by lying in the sand. (Details are at savethebeach.org.)
>
> I hope many of our neighbors will join in this day of action. It seems to me like the least we can do to help make sure our kids can enjoy the same California that's meant so much to us.

You don't even need a news story about global warming as a hook for your letter.

> Dear Editor:
>
> It saddens me how often we hear news about the polarization of our politics between left and right—I think our country has gotten stuck in a rut, and it's making it hard to deal with our most important problems, like global warming.

Citizens from the North Shore of Massachusetts gathered on Crane Beach in Ipswich. Organizer Franz Ingelfinger wrote a powerful letter to the editor of the *Boston Globe* to get word out about the event. (PHOTOGRAPH BY MIKE JOHNSON)

That's why I'm so glad that religious groups from around the county will be gathering for a special interfaith march to call for action on climate change. This is an issue that all people who care about God's creation can agree on, from evangelical Christians to Unitarians, from Jews to Muslims. We'll be walking a circuit of the seven houses of worship in downtown Jonesboro, stopping in each to offer prayers—and to replace the lightbulbs in each sanctuary with new compact fluorescent lights that symbolize our hope for a new and brighter energy future.

Our eighth and final stop will be at Congressman Johnson's district office, where we'll deliver a petition asking for action from Washington to help in this effort. All are welcome to join—we'll meet at the Lutheran Church at 10 AM on Saturday, April 6. Complete details can be found at www.circlethechurches.org.

Religious people can go on looking for things to disagree about—there will always be plenty. Or they can concentrate on the things they have in common, namely a reverence for the world God has made.

Basically, your letter to the editor should convey the core of the idea in a sentence and then provide enough details that interested readers can easily follow up. Keep it short because, if you don't, either the paper will cut the text of your letter for you or it won't find room to print it.

Most newspapers also let readers, especially those with some special expertise, write op-ed columns. These are longer than letters to the editor—seven hundred words is a standard length—and often they play off some event in the news. If there's a new scientific report on melting ice, say, that makes the newspapers, that's a good chance for one of your team to write an op-ed. It would likely summarize the news in a paragraph, and then go on to talk about what it might mean locally, before focusing on the chance that local residents have to do something about it by joining your campaign. Op-eds usually follow a form: one or two sentences to each paragraph, and the sentences should be pretty straightforward and punchy. If a member of your organizing team is a local expert on an issue, that person will have the best success rate placing a piece on the op-ed page. Your local public radio station may also feature listener commentaries that serve the same purpose. If there's a radio call-in show on a related topic, it might pay off to get a listener comment on the air that way. Finally, your local commercial radio stations may have morning drive-time shows during which the hosts conduct five- to ten-minute interviews with local residents. For these programs, it helps if the person can wield a sense of humor.

CREATE ADVANCE COVERAGE

If you're getting together to make signs for your demonstration, have your official spokesperson invite his or her closest media contacts, and let the reporters and editors know in advance that there will be good visual images as you're prepping for the big event. Or plan "prequel" or "teaser" events for your action. One idea is to deliver invitations to each of the members of your city council the week before. Tell the assignment editor the date and time when you'll be making the delivery, and add that you'll be giving each of the council members a compact fluorescent lightbulb at the same time. The basic principle is that journalists often like to cover things *before* they happen, whether you call it a scoop or call it good community service. And you like that, too, because it builds momentum for your event.

By the same token, you need to acknowledge the last-minute nature of the press. Some reporters may feel that if they cover the pre-event, there is nothing new to report once the main action arrives, and they may leave you wondering until the last minute about whether you'll get any coverage at all. We had been working on Step It Up for three months and had organized more than a thousand rallies, but we had gotten little national press coverage. Still, we didn't despair—we trusted that the media would come right before (and on) April 14. Sure enough, on the Friday before our big Saturday action, the press kicked in. A big Associated Press story went out on the national wire, and then TV and radio called, leading to appearances on ABC's *Good Morning America*, NPR's *Talk of the Nation*, and PBS's *News Hour with Jim Lehrer*. By the end of that Friday, there were hundreds of articles on Google News about Step It Up, a number that increased by an order of magnitude the next day when the actual events took place around the country. If you have cultivated your relationships and done your homework, it's reasonable to expect a payoff at the last minute, which is when journalism (the ultimate last-minute profession) bites.

BUILD LAST-MINUTE BUZZ

But count on nothing. Instead, you need to work the media right to the last minute. In the final twenty-four hours before your event, you want to create an overwhelming sense of urgency around it in the newsrooms.

Don't worry any longer about a single point of contact—obnoxiousness is no longer a big issue. Have six, eight, ten people call the news tips line for each of the papers and TV and radio stations the day before and the day of your event. The callers don't need to identify themselves as part of the planning team—they can just be citizens who have heard of a big event that's taking place the next day. Or they were just downtown and saw two hundred people gathering on bicycles and wanted to let the station know.

Precisely because news is a last-minute business, journalists are set to cover things on the spur of the moment, and you want to provide that last spur to get them into action. Sometimes last minute really means last minute. When we did our march across Vermont in 2006, we were coming into Burlington, a huge line of people stretching along the state highway. There were lots of cameras around, but no one from the Associated Press (which, as mentioned above, is one of the most important news outlets in any area). So Bill got on the cell phone as we walked and called the assignment desk. They were apologetic—short-staffed, couldn't leave the office. He described the scene and they conducted a short interview and, sure enough, within an hour, the story was out on the wire.

SPURRING THE MEDIA

Carry a simple media contact phone and e-mail list on the day of your action so that you can make calls to get spur-of-the-moment coverage.

PRACTICAL TIP

- Have your official media contact prewrite text messages and add media e-mail addresses to cell phones so that he or she can send an e-mail blast to the list half an hour before your event and right as the event kicks off or breaks news.

• Ask members of your organizing team, including anyone staffing a media check-in table at your event, to add key press phone numbers to their cell phone address books and quick-key contacts before the day of the event.

While you're following all this good advice, don't forget to take some time to make your own media. We live in an age less dependent on the formal press than any in American history—if you have a computer, you have a printing press of sorts. Use it, too.

MAKE IT LAST

As we finish this book, we're more aware than ever of how many things we have left uncovered—largely because we don't know much about them. It's our great hope that organizers will eventually move beyond the simple kinds of actions we've outlined here and explore the wide range of ways that citizens have figured out over the years to make change happen. The list of tactics runs from teach-ins and lectures to street theater, from direct actions like organizing lightbulb exchanges to sitting-in at your senator's office to blocking the tracks of the coal trains that daily crisscross the nation. You can run insurgent candidates for office or organize lawsuits, vigils, and fasts; you can use your power as a shareholder or a taxpayer. The river of nonviolent and creative action runs wide and deep. In the Resources section at the back of this book as well as the one on the Step It Up Web site, we have tried to point you toward a few of the many places you can read and learn more. We do not plan to organize a Step It Up day of action every spring—instead we're following our own advice and planning quick, ad hoc events to get the message out and keep the pressure on, including another nationwide day of action in November 2007.

Our goal in this book has been as much to suggest a mood as a plan, to talk about the ways that an era too often seen as apathetic might use the tools and the insights peculiar to our time to fight for a change that is essential for our future. This isn't a onetime event or an annual event; climate change is a crisis and it requires a new way of living our lives, for centuries to come. Yet if we had to list the main ideas that will bring together this movement we so need, they would be quite simple.

- **Take action.** Do *something*. Get together with whomever you can and brainstorm. All you need is a big piece of paper and an open mind. You don't have to come up with the perfect action, just something that excites you and that, by stretching, you think you can pull off. Plan it, do it, learn from it, repeat.

- **Don't fret about structure.** Far more than we need new organizations, we need nimble, relevant, strategic, and often temporary groups of people who can come together to do what needs to be done at the moment—and then do it again, with a whole different bunch of people, a few months later.

- **Emphasize openness.** Let people know what you're up to, invite them in to help, make them leaders. Figure out what you can work together on, not what divides you.

- **Go deep.** Without getting all sappy, we feel pretty strongly that silence, singing, deep bonds, and prayer (for those wired that way) make strong action in the face of real peril possible.

- **Have fun.** The best antidote for fear and powerlessness is joy. If you need to have meetings, make them fun—that is, full of good music and good food—if you can. Joy should not be postponed until after we have conquered global warming; it's precisely the fuel that will keep your passion burning for the long run.

We wish we could promise you that if enough people got involved we would solve the problem, but we can't—global warming is the biggest trouble humans have ever made for themselves, and its momentum has been snowballing without a response for a long time. We probably can't stop it altogether, and even if we could the world has myriad other sadnesses that need attention, too, from the gross inequality between peoples and nations to the scourge of war to the sheer loneliness that marks too much of modern life. Yet by getting deeply involved in fighting climate change, each of us can make the planet healthier and more livable and help create the kind of communities that can ward off the worst of what's coming—and also help us cope, together, with what we can't prevent.

We know one thing for certain—we've never enjoyed a year as much as the one we've just spent, very hard at work, with each other. To be able to move the world, if only a little, and to do it with friends old and new, is a great privilege. Let us know how it goes with you—e-mail us at organizers@stepitup2007 .org.

—Bill, Phil, Will, May, Jamie, Jeremy, and Jon,
the team at stepitup2007.org

RESOURCES

For additional sample materials from Step It Up 2007 and other supplemental online resources, visit the *Fight Global Warming Now* Web site at billmckibben.com/fightglobalwarmingnow. The site includes links to the Step It Up 2007 blog archive and nearly 1,000 site-by-site reports from the 1,400 events held on April 14, 2007.

BOOKS AND RESEARCH REPORTS

Alinksy, Saul. *Rules for Radicals: A Practical Primer for Realistic Radicals.* New York: Random House, 1971. The classic handbook on activism.

Bread and Puppet Theater. *Why Cheap Art? Manifesto.* Glover, Vt.: Bread and Puppet Press, n.d. A quick-and-dirty guide to the best of creative activism. The Bread and Puppet Press offers a range of materials on everything from how to make puppets to inspiring scribbles. You can view their catalog at breadandpuppet.org .

David, Laurie. *Stop Global Warming: The Solution Is You.* Golden, Colo.: Fulcrum Press, 2006. An environmental activist provides testimony of her grassroots efforts to stop global warming and shows how and why others can get personally involved.

De Rothschild, David. *The Live Earth Global Warming Survival Hand-book: 77 Essential Skills to Stop Climate Change.* Emmaus, Pa.: Rodale Press, 2007. The official companion volume to the July 2007 Live Earth concerts presents seventy-seven personal life choices anyone can make to help stop climate change.

Epstein, Paul R., and Evan Mills, eds. "Climate Change Futures: Health, Ecological and Economic Dimensions." Cambridge, Mass.: Center for Health and Global Environment, Harvard Medical School, with sponsorship from Swiss Re and the UN Development Programme, November 2005. Available at: swissre.com/INTERNET/pwswpspr.nsf/fmBookMarkFrameSet?ReadForm&BM=../vwAllby IDKeyLu/fstn-6hrreq?OpenDocument.

Gladwell, Malcolm. *The Tipping Point: How Little Things Can Make a Big Difference.* Boston: Little, Brown, 2000. A wonderful overview of creative ways to make your ideas "stick" and "tip."

Hawken, Paul. *Blessed Unrest: How the Largest Movement in the World Came into Being and Why No One Saw It Coming.* New York: Viking, 2007. An inspiring look at today's environmental movement.

Heath, Chip, and Dan Heath. *Made to Stick.* New York: Random House, 2007. A college professor and a corporate education researcher explain what has made some products, ideas, myths, trends, and movements in history catch on and become a part of the public conscience.

Hudema, Mike, and Jacob Rolfe. *An Action a Day Keeps Global Capitalism Away.* Toronto, Ont.: Between the Lines Press, 2004. For concerned citizens, an introduction to a variety of strategies, including social action, organizing, civil disobedience, and using the media.

Intergovernmental Panel on Climate Change. "Climate Change 2007: The Physical Science Basis," February 2007, www.ipcc.ch/SPM2 feb07.pdf. An essential report from the working group tasked with looking at the physical evidence for global warming in advance of the IPCC's five-year report, released in May. A number of technical and policy papers are available on the IPCC's Web site, www.ipcc.ch. Printed versions are available from Cambridge University Press.

Isham, Jonathan, and Sissel Waagy, ed. *Ignition: What You Can Do to Fight Global Warming and Spark a Movement*. Washington, D.C.: Island Press, 2007. Combining incisive essays with success stories and Web resources, this essential guide takes the global warming fight beyond changing lightbulbs and carpooling.

Lakoff, George. *Don't Think of an Elephant: Know Your Values and Frame the Debate*. White River Junction, Vt.: Chelsea Green Publishing, 2004. The definitive handbook on how to communicate effectively about key issues. Lakoff breaks down the ways in which conservatives have framed issues in the past and provides examples for progressives who wish to reframe the debate in their own terms.

Moyer, Bill et al. *Doing Democracy: The MAP Model for Organizing Social Movements*. Gabriola Island, B.C.: New Society Publishers, 2001. Provides both a theory and a working model for understanding and analyzing social movements, ensuring that they are successful in the long term. *Doing Democracy* outlines the eight stages of social movements, the four roles of activists, and case studies from numerous movements.

MAGAZINES AND PERIODICALS

The Adirondack Explorer (adirondackexplorer.org). A not-for-profit bimonthly magazine devoted to the East's great Adirondacks park and how to use it wisely, enjoy it fully, and protect it permanently.

Creation Care (creationcare.org). Evangelical Environmental Network's flagship publication, it provides biblically informed and timely articles on topics related to religion and environmental issues.

Orion (orionmagazine.org). One of the longest-standing nature-related publications, with a new emphasis on environmental issues.

Plenty (plentymag.com). An environmental media company dedicated to exploring and giving voice to the green revolution.

Yes! (yesmagazine.org). Supports the creation of a more just, sustainable, and compassionate world.

MOVIES

Everything's Cool. Film. Dir. Dan Gold and Judith Helfand. Wilmington, N.C., and New York: Working Films, 2007. An entertaining look at how Americans finally "got" global warming. See everythingscool.org for screening and DVD release dates.

The Great Warming. DVD. Dir. Michael Taylor. Montreal and Ottawa: Stonehaven Productions, 2007. Narrated by Alanis Morissette and Keanu Reeves, this dramatic film sweeps around the world to reveal how the changing climate is affecting people's lives. Four educational versions of the DVD are available for purchase at thegreatwarming.com.

An Inconvenient Truth. DVD. Dir. Davis Guggenheim. Hollywood, Calif.: Paramount Pictures, 2006. This riveting award-winning documentary presents Al Gore's famous PowerPoint lecture on the consequences of global warming and the myths and misconceptions that surround the science.

Marching for Action on Climate Change: Five Days Across Vermont with Bill McKibben and Friends. DVD. Dir. Jan Cannon. Charlotte, Vt.: Jan Cannon Films, 2007. A short, inspirational documentary of the walk across Vermont that features eloquent speeches about global warming and organizers strategizing their grassroots actions. Available at jancannonfilms.com/climatechange.htm.

CLIMATE CHANGE ORGANIZATIONS AND WEB SITES

Alliance for Climate Protection (allianceforclimateprotection.org). A nonpartisan organization addressing climate change through government, business (including energy, technology, finance, and media), and civil society groups.

Campus Climate Challenge (climatechallenge.org). A project of more than thirty leading youth organizations throughout the United States and Canada that helps students on college campuses and high schools to win 100% Clean Energy policies at their schools.

Chesapeake Climate Action Network (chesapeakeclimate.org). The first grassroots not-for-profit organization dedicated exclusively to fighting global warming in Maryland, Virginia, and Washington DC.

Climate Action Network (CAN) (climatenetwork.org). A worldwide network of nongovernmental organizations (NGOs) working to promote government and individual action to limit human-induced climate change to ecologically sustainable levels.

Climate Ark (climateark.org). A climate change and global warming portal and search engine that promotes public policy on global climate change through reductions in carbon dioxide and other emissions, renewable energy, energy conservation, and ending deforestation.

Climate Corps (climatecorps.org). An outgrowth of the Climate Project, a free membership organization devoted to slowing and reversing the current trend of global warming, which threatens the planet with massive and dire ecological and economic consequences if action is not taken immediately.

Climate Counts (climatecounts.org). A not-for-profit campaign that annually scores companies on the basis of their voluntary actions to reverse climate change. A unique collaboration between Stonyfield Farm and Clean Air–Cool Planet, Climate Counts helps people make critical distinctions between well-known brands that are leaders in the green movement.

Climate Crisis Coalition (climatecrisiscoalition.org). Seeks to broaden the circle of individuals, organizations, and constituencies engaged in the global warming issue, to link it with other issues, and to provide a structure to forge a common agenda and advance action plans with a united front.

Climate Project (theclimateproject.org). A not-for-profit organization dedicated to educating the public and governments on the growing crisis of global warming.

Climate USA (kyotoandbeyond.org/climateUSA.html). Focuses on efforts to create "A Call to Action to Midterm Congressional Candidates."

Climate Voters (climatevoters.org). A not-for-profit grassroots effort dedicated to making the prevention of catastrophic climate change one of America's top political priorities in the upcoming elections by 1) leveraging the voting power of individuals in this country who recognize the critical importance of addressing climate change; and 2) empowering this constituency to unite and demand leadership on climate change from their elected officials.

Cool Cities (coolcities.us). The Sierra Club's guide to the American cities that have signed the U.S. Mayors Climate Protection Agreement to stop global warming.

Exxpose Exxon (exxposeexxon.com). A collaborative effort of several of the nation's largest environmental and public advocacy organizations aimed at educating the public about ExxonMobil's efforts to block action on global warming, drill in the Arctic Refuge, and keep America addicted to oil.

Focus the Nation (focusthenation.org). Coordinating teams of faculty and students at over a thousand colleges, universities, and K-12 schools in the United States to collaboratively engage in a nationwide, interdisciplinary discussion about "Global Warming Solutions for America."

Global Warming Education Network (gwenet.org). A U.S.-based, not-for-profit organization dedicated to spreading awareness and encouraging action relating to global warming and its adverse impacts.

Heat Is On (heatison.org). A project to make global warming a priority issue during the presidential primaries. Educates voters, works with concerned political donors, and challenges the media to make climate change a top issue.

Home International (homeintl.org). Providing continuing community education and events about climate change for the student and philanthropic segments of our society to stimulate remedial and adaptive solutions by individuals, business, and government.

Intergovernmental Panel on Climate Change (IPCC) (www.ipcc.ch). Established by WMO and the UNEP to assess scientific, technical, and socioeconomic information relevant for the understanding of climate change, its potential impacts, and options for adaptation and mitigation.

It's Getting Hot in Here (itsgettinghotinhere.org). Dispatches from the global youth climate movement.

Low Carbon Diet Initiative (empowermentinstitute.net). Offers a suite of carbon-reduction tools centering on the *Low-Carbon Diet* workbook—an easy-to-use, illustrated guide that walks individuals or small groups (EcoTeams) through a time-tested CO_2-reduction program. Global Warming Café is a four-hour discussion-format workshop that anyone can host as a way to rally their community in response to the climate crisis. The Cool Community Campaign helps communities achieve a 20 percent reduction in carbon by 2010.

Medical Alliance to Stop Global Warming (psr.org). A new medical student/physician initiative to highlight health consequences of global warming and encourage action in the medical community.

Muse Campaign (musecampaign.org). A project of "Cool Our Planet," a not-for-profit organization. Links the world's musical and artistic talent with the business capability and community-building power of the Internet to educate the public, change individual lifestyles, and influence public policy on climate change. Also raises funds for projects that lower greenhouse gases and help victims of climate crisis.

Pew Center on Global Climate Change (pewclimate.org). A nonpartisan information source on global warming. The Pew Center's Business Environmental Leadership Council (BELC) is the largest U.S.-based business association focused on global warming solutions.

Play-It-Forward (play-it-forward.org). A movement based on the Oscar-winning documentary film *An Inconvenient Truth* and committed to inspiring people to take action to reverse the effects of global warming. Distributes free copies of *An Inconvenient Truth* for educational use.

Southern Alliance for Clean Energy (cleanenergy.org). A not-for-profit nonpartisan organization that promotes responsible energy choices that can help ensure clean, safe, and healthy communities throughout the Southeast United States.

Stop Global Warming (stopglobalwarming.org). A nonpartisan effort to bring citizens together to declare that global warming is here now and that it is time to demand solutions.

Students United for a Responsible Global Environment (SURGE) (surgenetwork.org). A North Carolina–based student network that operates a moderated listserv of national announcements on climate change and other environmental issues.

United Nations Framework Convention on Climate Change (unfccc.int/2860.php). Over a decade ago, most countries joined an international treaty—the United Nations Framework Convention on Climate Change (UNFCCC)—to consider what could be done to reduce global warming. This site contains introductory and in-depth publications, the official UNFCCC and Kyoto Protocol texts, and a search engine to the UNFCCC library.

United States Conference of Mayors Climate Protection Page (usmayors.org/climateprotection). Background information on the U.S. Mayors Climate Protection Agreement, links to mayors making news on global warming, and a list or resolutions adopted by the conference since 2000.

U.S. Climate Emergency Council (climateemergency.org). A not-for-profit organization dedicated to rigorous grassroots action in the fight to stop global warming and promote a clean energy future.

ENVIRONMENTAL ORGANIZATIONS AND WEB SITES

Adirondack Mountain Club (adk.org). A not-for-profit membership organization that protects wildlands and waters through a balanced approach of conservation and advocacy, environmental education, and responsible recreation.

Appalachian Voices (appalachianvoices.org). Brings people together to solve the environmental problems having the greatest impact on the central and southern Appalachian Mountains.

Bioneers (bioneers.org). A not-for-profit organization that promotes practical environmental solutions and innovative social strategies for restoring the earth and communities.

Blue Frontier Campaign (bluefront.org). Works to support seaweed (marine grassroots) efforts at the local, regional, and national levels, with an emphasis on bottom-up organizing to bring the voice of citizen activists into national decision making.

Center for Biological Diversity (biologicaldiversity.org). A conservation organization dedicated to protecting endangered plants, animals, and the wild places we all need to thrive.

Earth Day Network (earthday.net). Founded by the organizers of the first Earth Day in 1970, it promotes environmental citizenship and year-round progressive action worldwide.

Environmental News Network (enn.com). Comprehensive news and commentary about living on the earth. Readers can sign up for an RSS feed of the site or individual "channels," including one on global warming and climate change news.

Friends of the Earth (foe.org). Defends the planet and champions a healthy and just world. Active in seventy countries, Friends of the

Earth is the world's largest network of grassroots environmental groups.

Greenpeace (greenpeace.org). Focuses on the most crucial worldwide threats to our planet's biodiversity and environment.

Green Seniors (greenseniors.org). Environmental action with no age limit.

Grist (grist.org). The best in online environmental journalism: "Doom and gloom with a sense of humor."

Gulf Restoration Network (healthygulf.org). A diverse network of local, regional, and national groups dedicated to protecting and restoring the valuable natural resources of the Gulf of Mexico. The GRN has members in the five Gulf states of Texas, Louisiana, Mississippi, Alabama, and Florida and nationwide.

Honor the Earth (honorearth.org). Supports Native environmental issues and develops needed financial and political resources for the survival of sustainable Native communities.

League of Conservation Voters (LCV) (lcv.org). The independent political voice for the environment. To secure the environmental future of our planet and advocate for sound environmental policies and the election of pro-environmental candidates who will adopt and implement such policies.

Mother Earth News (motherearthnews.com). The original guide to living wisely.

National Wildlife Federation (nwf.org). Inspires Americans to protect wildlife for our children's future.

Natural Resources Defense Council (nrdc.org). Uses law, science, and the support of 1.2 million members and online activists to protect the planet's wildlife and wild places and ensure a safe and healthy environment for all living things.

Northwest Earth Institute (nwei.org). Offers programs that encourage participants to explore their values, attitudes, and actions through discussion with other people.

Orion Grassroots Network (orionsociety.org). The fastest-growing hub of environmental and community organizations in North America, actively supporting more than a thousand organizations.

Rainforest Action Network (ran.org). Works to protect the earth and support human rights through corporate campaigning, education, grassroots organizing, and nonviolent direct action.

Reef Relief (reefrelief.org). A not-for-profit grassroots membership organization dedicated to preserving and protecting living coral reef ecosystems through local, regional, and international efforts.

Student Conservation Voters (lcvef.org/student-conservation-voters). A national youth program of the League of Conservation Voters Education Fund that trains and organizes students to make their voices heard in elections and with elected officials.

Urban Divers Estuary Conservancy (urbandivers.org). A not-for-profit grassroots environmental organization composed of volunteer scientific divers, waterway stewards, and citizen monitors committed to active participation in the public education, restoration, conservation, and protection of our rivers, oceans, and marine wildlife, with a special focus on restoration of the New York–New Jersey Harbor Estuary.

What's in My Water (whatsinmywater.com). Educating the public on the EPA's role to enforce water-pollution regulations.

World Resources Institute (www.wri.org/climate). An environmental think tank that works to reverse rapid degradation of ecosystems; protect the global climate system from harm due to emissions of greenhouse gases; harness markets and enterprises to expand economic opportunity; and protect the environment.

GREEN-ORIENTED BUSINESS AND SUSTAINABLE ENERGY ORGANIZATIONS AND WEB SITES

1% for the Planet (onepercentfortheplanet.org). An alliance of companies that recognize the true cost of doing business and donate

1 percent of their sales to environmental organizations worldwide.

Alliance for a Sustainable Future (asustainablefuture.org). An organization committed to bringing to life the practices, stories, and images that can help generate a public-wide shift to sustainable living.

American Solar Energy Society (ASES) (ases.org). The U.S. section of the International Solar Energy Society, a not-for-profit organization dedicated to the development and adoption of renewable energy in all its forms, including solar energy, wind energy, geothermal energy, hydrogen energy, ocean energy, biofuels energy, and energy efficiency.

Better World Club (betterworldclub.com). The nation's only environmentally friendly auto club. BWC donates 1 percent of revenues to environmental cleanup and advocacy, and offers unique, eco-friendly roadside assistance, insurance, and travel alternatives.

Co-Op America (coopamerica.org). To harness economic power—the strength of consumers, investors, businesses, and the marketplace—to create a socially just and environmentally sustainable society.

ECOSA Institute (ecosainstitute.org). Offers semester programs in sustainable design and summer workshops in permaculture and alternative construction.

Green Banners (greenbanners.com). Offers photo-quality, full-color printing for a wide variety of banner and signage on recycled, renewable, and recyclable earth-friendly materials.

Green Exchange (greenexchange.com). The first business community committed to environmental sustainability, profit, and positive social impact. A 250,000-square-foot concrete loft building showcases a mix of complementary businesses that offers a unique collection of leading-edge products and services to the environmentally conscientious consumer.

Imagining Tomorrow (itomorrow.theforesightproject.org). A creative writing and video contest about clean energy for all high school students in the United States, with state and regional levels. Top entries go on to the national level, with ten thousand dollars committed in prizes to the national winners.

International Forum for Globalization (ifg.org). An alliance of sixty leading activists, scholars, economists, researchers, and writers formed to stimulate new thinking, joint activity, and public education in response to economic globalization.

Kitchen Gardeners International (kitchengardeners.org). Empowering individuals, families, and communities to achieve greater levels of food self-reliance through the promotion of kitchen gardening, home cooking, and sustainable local food systems.

Mesa Environmental Sciences, Inc. (mesasolar.com). A woman-owned energy and environmental services company.

Native Energy (nativeenergy.com). A privately held Native American energy company that helps build Native, farmer-owned, community-based renewable energy projects.

Natural Capitalism Solutions (natcapsolutions.org). Educates senior decision makers in business, government, and civil society about the principles of sustainability.

New American Dream (newdream.org). Helps Americans consume responsibly to protect the environment, enhance quality of life, and promote social justice.

Northeast Sustainable Energy Association (nesea.org). The Northeast's leading organization of professionals and concerned citizens working in sustainable energy and whole-systems thinking.

Oil Change International (priceofoil.org). A campaign to expose the true costs of oil and facilitate the coming transition toward clean energy.

Organic Home (organichomedesign.com). Offers clients a unique home and office service providing green cleaning, organizing, design, and consultation.

Project Laundry List (laundrylist.org). Uses words, images, and advocacy to educate people about how simple lifestyle modifications, including air-drying clothes, reduce our dependence on environmentally and culturally costly energy sources. The group hosts National Hanging Out Day events on April 19 each year.

Real Green Goods (realgreengoods.com). A small retail operation offering the highest standards in eco-friendly products with the lowest impact to Earth and its inhabitants. Dedicated to selling organic products, and Fair Trade and U.S.-made items.

Renewable Energy Trust (masstech.org/renewableenergy/index.html). Seeks to maximize environmental and economic benefits for citizens by pioneering and promoting clean energy technologies and fostering the emergence of sustainable markets for electricity generated from renewable sources.

SEPTA (septa.org). The fifth-largest transit agency nationwide, providing more than 1.1 million daily trips to residents of southeastern Pennsylvania, northern Delaware, and Trenton, New Jersey. Fleet consists of hybrid-electric buses, clean diesel buses, trolleys, vans, and all-electric subway-elevated and regional rail trains. As a result, the Philadelphia region has the second-lowest ratio of cars per capita of any major U.S. metropolis, resulting in less highway congestion and the opportunity to enjoy cleaner air. All buses contain bicycle racks.

Solar Energy Industries Association (seia.org). The national trade association of solar energy manufacturers, dealers, distributors, contractors, installers, architects, consultants, and marketers. Works to expand the use of solar technologies in the global marketplace.

Stonyfield Farm (stonyfieldfarm.com). The leading producer of organic yogurt in the world and a leader in social and environmental responsibility in business, Stonyfield Farm gives away 10 percent of its profits to organizations and projects that work to protect and restore the earth.

Vote Solar Initiative (votesolar.org). An organization dedicated to bringing solar energy into the mainstream.

WaterCheck.biz (watercheck.biz). A leading marketer of water-related products that give people a new experience of their H_2O. The site is also the home of the famous WaterManBlog, which keeps you up to date about all things H_2O. A portion of all revenues supports water-sustainability programs throughout the globe.

Working Assets (workingassets.org). A business established in 1985 with an environmentally sustainable outlook. When customers use a Working Assets donation-linked service (long distance, wireless, and credit card), the company donates a portion of the charges to nonprofit progressive causes.

Yestermorrow (yestermorrow.org). A design/build school dedicated to a better, more sustainable world. Provides hands-on education that integrates design and construction as a creative, interactive process.

POLITICAL ORGANIZATIONS AND WEB SITES

Alternet (alternet.org). An award-winning online news magazine and community that creates original journalism and amplifies dozens of other independent media sources.

Avaaz.org (avaaz.org/en). A community of global citizens who take action on the major issues facing the world today based on a vision of globalization that has a human face.

Bus Project (busproject.org). A grassroots movement that has mobilized thousands of volunteers and activists in the state of Oregon. It creates forums to learn about candidates and policy issues, and "Bus-PAC" works to elect the best progressive candidates in the state.

COA News (coanews.org). A not-for-profit online news network featuring diverse, credible independent news and current affairs.

Codepink—Women for Peace (codepinkalert.org). A women-initiated grassroots peace and social justice movement working to end the war in Iraq, stop new wars, and redirect our resources into health care, education, and other life-affirming activities.

Daily Kos (dailykos.com). News from the progressive world.

Friends of Animals (friendsofanimals.org). An international, not-for-profit organizational base for the animal rights movement in New York, established in 1957. FoA brings critical habitat and environmental issues into the animal-advocacy sphere, as its advocates for the right of animals to live free according to their own terms.

Global Exchange (globalexchange.org). A membership-based international human rights organization dedicated to promoting social, economic, and environmental justice around the world.

Hello Cool World (hellocoolworld.com). Makes connections and supports and generates social cause campaigns. Notorious for its on-going association with the blockbuster film *The Corporation*. Both cheeky and smart.

MoveOn.org (moveon.org). Brings Americans back into the political process. With over 3.3 million members across America—from carpenters to stay-at-home moms to business leaders—MoveOn works to realize the progressive promise of our country.

No War No Warming! (nowarnowarming.org). Acknowledges the ways in which war and climate change are linked and the need for people throughout the world to take action to end both.

Physicians for Social Responsibility (psr.org). A not-for-profit advocacy organization that is the medical and public health voice for policies to stop nuclear war and proliferation and to slow and reverse global warming and toxic degradation of the environment.

Service Employees International Union (SEIU) (seiu.org). The fastest-growing union in North America. SEIU members are winning better wages, health care, and more secure jobs for their communities while uniting their strength with their counterparts around the world to help ensure that workers, not just corporations and CEOs, benefit from the global economy.

Topics Education (topicseducation.com). Founded in 1995 by a former teacher, this is a values-led education outreach and communications company based in Charlotte, North Carolina.

TrueMajority (truemajority.org). Monitors Washington and sends short e-mail alerts for action on specific issues.

United for Peace and Justice (unitedforpeace.org). A coalition of more than 1,300 local and national groups throughout the United States that have joined to protest the Iraq war and oppose the U.S. government's policy of permanent warfare and empire building.

U.S. PIRG (uspirg.org). A federation of state Public Interest Research Groups (PIRGs) with a strong network of researchers, advocates, organizers, and students that stands up to powerful special interests on issues to promote clean air and water, protect open space, stop identity theft, fight political corruption, provide safe and affordable prescription drugs, and strengthen voting rights.

Visual Resistance (visualresistance.org). A crew of artists and activists based in Brooklyn who first came together to organize the "No RNC Poster Project" in summer 2004. Uses art to transform and liberate public space, addresses local struggles, urban development, freedom of speech, and political repression. The Web site contains a blog, instructions on how to make street art, and information on upcoming events.

Vote Smart (votesmart.org/index). A nonpartisan not-for-profit voter education project that includes a Web-based database of biographical and contact information for elected officials at the local, state, and federal levels. Includes data on voting records and state ballot initiatives.

RELIGIOUS ORGANIZATIONS AND WEB SITES

Coalition on the Environment and Jewish Life (coejl.org). Seeks to expand the contemporary understanding of such Jewish values as *tikkun olam* (repairing the world) and *tzedek* (justice) to include the protection of both people and other species from environmental degradation.

Earth Ministry (earthministry.org). Inspires and mobilizes the Christian community to play a leadership role in building a just and sustainable future.

Eco-Justice Ministries (eco-justice.org). An independent, ecumenical agency that helps churches answer the call to care for all of God's creation and develop ministries that are faithful, relevant, and effective in working toward social justice and environmental sustainability.

Evangelical Environmental Network (creationcare.org). An evangelical ministry dedicated to the environment through the teachings of the Bible to "declare the Lordship of Christ over all creation" (Col. 1:15–20; Jn 1:1–4; Heb. 1:2–3).

Interfaith Walk for Climate Rescue (climatewalk.org). An archive of materials from the March 2007 walk from Northampton to Boston, Massachusetts, organized and composed of religious leaders and activists demanding swift, bold, and comprehensive political action to address global warming.

Islamic Foundation for Ecology and Environment Sciences (ifees.org). An internationally recognized organization articulating the Islamic position on environmental issues. It creates and distributes teaching and training materials on the rapid destruction of ecosystems and manages an experimental center focusing on land use, organic farming, and the development of alternative energy technology.

National Council of Churches (nccecojustice.org). Works in cooperation with the NCC Eco-Justice Working Group to provide an opportunity for Protestant and Orthodox denominations to work together to protect and restore the environment.

National Religious Partnership for the Environment (nrpe.org). Guided by biblical teaching, seeks to encourage people of faith to weave values and programs of care for God's creation throughout the entire fabric of religious life.

Presbyterians for Restoring Creation (prcweb.org). A nationwide grassroots network that connects, equips, and inspires Presbyterians and other people of faith to care for God's creation.

Regeneration Project/Interfaith Power and Light (theregeneration project.org). A campaign to mobilize a national religious response to global warming while promoting renewable energy, energy efficiency, and conservation.

Web of Creation (webofcreation.org). A religious organization fostering the eco-justice movement. Also provides information and connections for theology students interested in environmental ministry.

SCIENCE WEB SITES

Environmental Protection Agency Glossary of Climate Change Terms (epa.gov/climatechange/glossary.html). An overview of key global warming terms from "aerosol" to "wastewater."

Nature **Climate Feedback** (blogs.nature.com/climatefeedback). An expert blog on the science and social implications of climate change from the renowned journal *Nature.*

New Scientist **Climate Change Special Report** (environment.newscientist .com/channel/earth/climate-change). A special report explaining the potential threats of global warming, the physics of the "greenhouse effect," and what must be done in the future.

Real Climate (realclimate.org). A commentary site on climate science by working climate scientists, free from politics and economic discussion. The site aims to provide a quick response to developing stories and provide the context sometimes missing in mainstream commentary.

Scientific American **Observations** (blog.sciam.com). A blog from the editors of *Scientific American* magazine that features links to articles, photos, and videos on global warming and climate change.

Seed **ScienceBlogs Planet Earth Channel** (scienceblogs.com/channel/planet-earth). The largest online community dedicated to science featuring leading bloggers from a wide array of scientific disciplines.

Union of Concerned Scientists (ucsusa.org). The leading science-based not-for-profit working for a healthy environment and a safer world. Combines independent scientific research and citizen action to develop innovative, practical solutions and to secure responsible changes in government policy, corporate practices, and consumer choices. Programs include the Renewable Electricity Standards Toolkit, a database of state regulations on clean energy, and numerous papers on personal and policy actions for stopping climate change.

GENERAL SERVICE WEB SITES

A full overview of Web tools and resources—from e-mail to photo sharing to RSS feeds—is included in chapter 6, "Make It Wired."

CaféPress (cafepress.com). A global and growing network of over 2.5 million independent shopkeepers and members who sell print-on-demand merchandise ranging from coffee mugs to T-shirts.

Craigslist (craigslist.org). A network of online city-based communities featuring free classified advertisements and forums sorted by topic. Craigslist is the leading classifieds service in any medium

Freecycle (freecycle.org). A grassroots and entirely nonprofit movement to further develop community reuse and recycling on a local scale in over seventy countries. Each group listserv is moderated by a local volunteer.

United States Newspaper and News Media Guide (abyznews.com/unite.htm). An online guide to newspaper and Internet news sources, listed by city and by state.

ACKNOWLEDGMENTS

To our parents, spouses, and children:

Andrei and Corina Aroneanu

Elizabeth and John Bates

Jill Hunting and Tim Boeve

Catherine and John Henn

Steve and Tina Osborn

Stan Warnow and Cathy Hiller

Peggy McKibben, Sue Halpern, and Sophie McKibben

**Through Middlebury and the greater
Vermont community:**

Robbie Adler

Corinne Almquist

Laura Cary

Liz Cunningham of Bioneers

Greg Dennis

John Elder

Thomas Hand of Native Energy

Erik Hoffner at *Orion*

Jon Isham

Tracy Himmel Isham

Mike Ives

Fielding Jenks

Nan Jenks-Jay

Connie Leach

Retta Leaphart

Ron Liebowitz

Mary Catherine McElroy

The Middlebury College Sunday Night Group

Joey Miller of Vermont Natural Resources Council

Julia Proctor

Michael Silberman of EchoDitto

Senator Bernie Sanders

Andrew Savage

Becca Sobel and the Greenpeace Walk Team

Peter Viola at *Orion*

Representative Peter Welch

Through the grassroots climate movement:

Meg Boyle

Courtney Fryxell and the League of Conservation Voters for
    putting us up

Ross Gelbspan

Ted Glick

Mike Hudema and the folks at Global Exchange

Mike Tidwell

Van Jones

Coalition partners and staff of the Energy Action Coalition, especially:

Arthur Coulston

Jared Duval

Josh Lynch

Billy Parish

Jessy Tolkan

Liz Veazey

Through Step It Up:

Kate Abend of the Union of Concerned Scientists USA

Harriet Barlow and the Blue Mountain Center

Chelsea Bassett of the DemocracyInAction.org team

Frances Beinecke

Cal Dewitt

Laurie David

The Gallerists

Chip Giller of Grist

Ellen Golombek at the Service Employees International Union

Eban Goodstein of Focus the Nation

Al Gore, for walking through the door with *An Inconvenient Truth*

Elysa Hammond of ClifBar

Paul Hawken

Judith Helfand and Dan Gold, directors of *Everything's Cool*, and everyone at Working Films

Ilyse Hogue and our friends at MoveOn.org

Shawnee Hoover of Exxpose Exxon

Lisa Hymas of Grist

Leslie Kagan and United for Peace and Justice

Kevin Knobloch of the Union of Concerned Scientists USA

Felix Kramer

Nancy Kricorian of Code Pink

Winona LaDuke

Hunter Lovins

Carl Pope, Odette Mucha, and the Sierra Club

The MUSE campaign

John Passacantando and Greenpeace

David, Margot, Seth, and the rest of the team at Radical Designs

Lisa Renstrom

Joel Rogers

Levana Saxon and Rainforest Action Network

Larry Schweiger

Michael Silberman

David Tuft and the Natural Resources Defense Council

Tim Walker and the BiroCreative team

The Step It Up organizers

The photographers who documented Step It Up

The donors who made Step It Up possible

**Through the Watkins Loomis Agency and
Henry Holt and Company**

Gloria Loomis and the whole team at Watkins Loomis

Nicholas Caruso

Robin Dennis

Paul Golob

Justin Golenbock

Tara Kennedy

Elyse Lipman

Chris O'Connell

Richard Rhorer

Maggie Richards

John Sterling

INDEX

Entries in *italics* refer to illustrations

ABOUT THE AUTHORS

BILL McKIBBEN is the author of ten books, including *The End of Nature* and *Deep Economy*. A former staff writer for *The New Yorker*, he writes regularly for *Harper's* and *The Atlantic Monthly*, among other publications.

THE STEP IT UP TEAM are activists Phil Aroneanu, Will Bates, May Boeve, Jamie Henn, Jeremy Osborn, and Jon Warnow. Step It Up, the national day of action on climate change, involved more than 1,400 events in all fifty states, the largest environmental protest in a generation.

Books by Bill McKibben available from
Holt Paperbacks

Deep Economy—In this powerful and provocative manifesto, Bill McKibben offers the biggest challenge in a generation to the prevailing view of our economy. For the first time in human history, he observes, "more" is no longer synonymous with "better"—indeed, for many of us, they have become almost opposites. McKibben puts forward a new way to think about the things we buy, the food we eat, the energy we use, and the money that pays for it all.

Enough—McKibben turns his eye to an array of technologies that could change our relationship not with the rest of nature but with ourselves. He explores the frontiers of genetic engineering, robotics, and nanotechnology—which we are approaching with astonishing speed—and shows that each threatens to take us past a point of no return. This wise and eloquent book argues that we cannot forever grow in reach and power—that we must at last learn how to say, "Enough."

Fight Global Warming Now—McKibben and the Step It Up team of organizers, who coordinated a national day of rallies on April 14, 2007, provide the facts of what must change to save the climate and show how to build the fight in your community, church, or college. They describe how to launch online grassroots campaigns, generate persuasive political pressure, plan high-profile events that will draw media attention, and other effective actions. This essential book offers the blueprint for a mighty new movement against the most urgent challenge facing us today.